SIGNAL AND IMAGE RESTORATION

Information-Theoretic Approaches

SIGNAL AND IMAGE RESTORATION
Information-Theoretic Approaches

Joseph Noonan and Prabahan Basu

SPIE
PRESS

Bellingham, Washington USA

Library of Congress Cataloging-in-Publication Data

Noonan, Joseph.
  Signal and image restoration : information-theoretic approaches / Joseph Noonan, Prabahan Basu.
      p. cm.
  Includes bibliographical references and index.
  ISBN 978-0-8194-8821-3
  1. Signal theory (Telecommunication) 2. Data recovery (Computer science) 3. Image reconstruction. I. Basu, Prabahan. II. Title.
   TK5102.92.N66 2011
   621.382'23--dc23
                                   2011046108
Published by

SPIE
P.O. Box 10
Bellingham, Washington 98227-0010 USA
Phone: +1 360.676.3290
Fax: +1 360.647.1445
Email: Books@spie.org
Web: http://spie.org

Printed in the United States of America.

First Printing

To my Family.

-   J.N.

To my father, J.P., my wife, Rituparna, and my mother, Rekha.

-   P.B.

Contents

Preface

The goal of this book is to present a unified information-theoretic approach to solve estimation problems commonly faced in signal-processing applications. We provide both new approaches to this problem as well as new interpretations for existing techniques. Major applications of this work include image restoration, communication channel estimation, text restoration, and system modeling. A general approach to solving a number of detection and estimation problems utilizing concepts from information theory is developed. The theoretical development of the approach as well as important applications is given. Additionally, the work of other researchers is shown to be special cases of this general technique.

Decision making in the context of detection and estimation problems requires assumptions about the probability model. The requirement for assumptions derives from the lack of sufficient information to arrive at the "right" solution. By making appropriate assumptions one can choose a single best estimate of the "right" solution from the available set of candidate solutions. In other words, these assumptions fill the gap, or resolve the uncertainty, that is left by the unavailability of the required information.

Information theory offers a powerful approach to the problem of assigning probability models. In Part II, we examine the use of information theory for probability modeling and establish estimates of the error bounds on the resulting models. In Part III, we consider the application of information-theoretic functional to the problem of estimation. Specifically, this research concerns the application of information-theoretic concepts to problems of deconvolution and system modeling. We develop a general iterative mapping derived using the mutual information functional to establish a criterion for optimality in solving problems of signal and image restoration and resolution enhancement. The mathematical structure is defined and the convergence of the mapping is established both in the general formulation and for particular examples. Some popular techniques for signal restoration, such as that proposed by Van Cittert and derived independently of information theory, are shown to be special cases of this general approach. Additionally, the approach allows the inclusion of prior partial knowledge of the desired solution and testing of various hypotheses concerning the solution characteristics. The work herein offers both new approaches to these problems as well as a theoretically sound general framework unifying the work of previous researchers.

Joseph Noonan and Prabahan Basu
Medford, MA
December 2011

Part I

Preliminaries

Chapter 1
Information Theory Preliminaries

1.1 Scope

The quality of any decision depends both upon the amount of relevant information available as well as the quality of the decision-making process itself. In this book, we investigate decision making in the context of detection and estimation problems that are statistically modeled according to some assumed probability distribution. Assumptions become necessary to help find *true* solutions when facing problems where available information is insufficient. Appropriate assumptions provide guidance when selecting the *best* solution from a given set of candidates.

However, new assumptions impart new information. Consequently, making the minimum number of assumptions consistent with available evidence is vital; making these assumptions using principled methods is also key. Given the important role of *information* in these problems, we appeal to information theory as a principled approach to making appropriate assumptions and assigning meaningful models.

Much of this book deals with the application of information-theoretic notions to problems of estimation. Following a brief review of information theory basics, we introduce the inverse problem and its role in signal restoration. Next, we review both classical and recent methods for solving the inverse problem. Part I of this monograph explores density-estimation techniques. We first review one such technique—maximum-entropy (ME) estimation. Then, we present a novel result which establishes confidence bounds on these estimates. In Part II, we introduce a framework for solving inverse problems to find unique solutions that maintain fidelity with observed data. We conclude the book with some examples and applications of the proposed algorithms to common problems encountered in signal processing.

1.2 History

The pioneers of modern information theory were the well-known scientists H. Nyquist, V. A. Kotelnikov, and R. V. L. Hartley. After World War II, Claude E.

Shannon and Norbert Wiener joined them in pushing the boundaries of our understanding of the concepts of information and probability. Early publications by Shannon and Wiener initiated a flurry of interest and activity in the related fields of information theory and probability.[1,2]

Wiener concerned himself with the important and difficult problem of extracting signals of a known ensemble for accurate and fast transmission, especially in the presence of noise. This problem is associated with the transmitting end of a communication system. In contrast, Shannon focused on the issue of the capacity of the channel itself. Shannon made the crucial observation that channel capacity by itself does not place a limit on the accuracy of transmission. Instead, it is the combination of channel capacity and transmission rate that together control the accuracy of communication. More generally, information theory as outlined by Shannon led to the discovery of theorems, functions, and techniques that helped to explain many existing *ad-hoc* encoding methods and techniques. It bears mentioning that while Wiener's work provided a constructive recipe for building optimal systems, Shannon's theory only offers theoretical bounds on the rate of information transmission and reception. In that sense, Shannon's theory is not constructive and does not offer any recipes for achieving these bounds.

While coding theory is probably the most well-known application of information theory, there have been broad efforts at applying information theory to other problems such as signal estimation and signal restoration. One component of the general signal-restoration problem is the assignment of a prior probability density function (PDF).

Until recently, Laplace's principle of insufficient reason held sway. The principle states that given a set of mutually exclusive and exhaustive events about which no probabilistic knowledge is available, we should assume all events to be equally likely. While the principle seems attractive at first blush, it hides the fact that assuming a uniform distribution over all outcomes is, in effect, an assumption about the probability model. Nevertheless, while the principle offers a reasonable course to follow in the total absence of any knowledge, it does not provide a mechanism for dealing with the case when some prior knowledge is available. Such prior knowledge usually renders some events more probable than others. The central problem considered in this book is the problem of assignment of probability densities when some prior information is available.

Stated explicitly, the following is the problem we wish to solve: there is some random phenomenon occurring that we wish to study. Associated with this phenomenon is some underlying probability model about which we have certain partial information, usually obtained by measurement. In order to solve any decision or estimation problems on this model, one must first determine the associated PDF. We therefore ask, "What density should we assign based on the partial information available?"

In developing a methodology for assigning densities, certain principles are self-evident. For example, it is clear that making an arbitrary guess does not serve any purpose. Also, ensuring that the density is consistent with whatever information is available is a good idea. We do not want to propose a model that starts its existence by denying the validity of observed data. Consistency is another desirable feature; given the same prior information and the same set of observations, the proposed scheme should result in the same estimate in every trial. While we are dealing with probabilistic entities, we do not want the model itself to exhibit unstable, probabilistic characteristics. The need for a principled approach is thus easily motivated. Other advantages that follow from a principled approach are that such an approach lends itself to mathematical analysis, thereby enabling further principled enhancements whose performance can be predicted and then validated. Finally, we emphasize again the need for minimalist assumptions. In plain English, the distinction is that between having a considered opinion and being opinionated. Our goal is to create a framework that allows the establishment of considered opinions while avoiding the pitfalls that come out of extreme, unsubstantiated opinions.

1.2.1 Background

In the discussion so far, we have outlined our goal of leveraging notions from information theory to establish probability models most faithful to the phenomenon under study. Surprisingly, the first work in this area was by Gibbs, who was using an entropy functional for the development of fundamental laws in statistical mechanics even before the idea of information theory itself was born. Once Shannon presented his mathematical theory of communications,[3] E.T. Jaynes (a fascinating and radical figure in the history of probabilistic thought),[4-6] recognized the connection between the early work of Gibbs and Shannon's information theory. Using the ideas of information theory, Jaynes then successfully applied entropy maximization to determine some of the laws of statistical thermodynamics.

Earlier, we mentioned Laplace's principle of insufficient reason based on which a uniform prior PDF is assigned in the face of complete uncertainty about probable outcomes. Our goal is to modify this principle to allow the principled inclusion of partial knowledge about the probable events. Jaynes made an important contribution[5,6] towards achieving this goal by suggesting that the assignment of priors should be such that the resulting distribution maximizes the entropy of the random variable, subject to constraints reflecting the known or available partial knowledge. Entropy maximization implies that one derives only the incontrovertible information that is contained in the partial knowledge. By attempting to maximize the uncertainty that is left even after the partial knowledge is provided, we raise our threshold for what constitutes acceptable information. A little reflection should convince the reader that this approach is the most conservative approach from an information-theoretic sense.

1.2.2 Definitions and concepts

The notion of entropy constitutes the fundamental building block of information theory. Entropy is a measure of the uncertainty about the outcome of a random process. Intuition tells us that if the distribution is a Dirac delta with a nonzero value for one outcome and zero for all others, then the entropy should be zero because there is no uncertainty about the outcome. Also, a uniform distribution over all probable outcomes does not help us identify a single-most probable outcome and therefore has the most entropy.

Based on four fundamental axioms or desiderata, researchers have derived the following functional form for information entropy:

$$H(x) = -\sum_x p(x) \log p(x), \tag{1.1}$$

where x is a discrete random variable and $p(x)$ is the probability mass function (PMF) of x. Clearly, it seems easy to extend the definition to the case of the continuous random variable by simply replacing the sum with an integral and the PMF with a PDF,

$$H(x) = -\int_x p(x) \log p(x), \tag{1.2}$$

where x is a continuous random variable and $p(x)$ is the PDF of x. In the case of a continuous random variable, the entropy functional listed above is also referred to by some authors as differential entropy. Differential entropy has the following properties:

1. $H(x + c) = H(x)$, if c is a constant.

2. $H(ax) = H(x) + \log |a|$, where a is a scaling constant.

3. $H(AX) = H(X) + \log(\det(A))$, where A is a matrix and X is a vector.

But, unlike the case of discrete random variables, the entropy functional for continuous random variables exhibits some undesirable characteristics:

1. The entropy may be negative.

2. The entropy may be infinitely large.

3. The entropy is not necessarily invariant under linear transformation of the random variable.

For obvious reasons, the above characteristics dilute the argument for the use of entropy as a measure of uncertainty in the case of continuous random variables. Fortunately, in most cases we are less interested in the entropy of any one variable and more in the relationship between random variables. In other words, for most applications our interest is in the correlation or predictive power of random variables, i.e., how much does random variable x tell me about random variable y? This is interesting because one is often unable to directly measure the phenomenon of interest, but indirect measurements are possible. For example, we cannot directly measure the temperature of the sun, but we can measure the amount of radiation that reaches the earth's surface. This concept also ties in neatly with our notion of partial knowledge—what partial knowledge of y does a measurement of x provide? In this context, researchers introduced the notion of mutual information (MI). The MI functional involves two random variables and measures, in an information-theoretic sense, the correlation between two random variables. The MI functional of two random variables, x and y, is expressed as

$$I(x, y) = H(x) - H(x|y). \tag{1.3}$$

In communications applications, it is common to think of x as the transmitted signal and y as the received signal. MI is then directly related to the capacity of the propagation channel: $H(x)$ is the entropy of the source, and $H(x|y)$ is the uncertainty about x that remains after y is known. Thus, $I(x, y)$ is the loss of information due to the channel distortion or noise. Channel capacity is one statistic that is useful in analyzing the efficiency of communication systems.[7,8] In any case, this view of the MI functional makes it clear that the MI functional is a useful tool for measuring system performance and provides us with a framework for characterizing the dependence between the input and output of a system. Another interesting characteristic of the MI functional is that $I(x, y)$ is the same as $I(y, x)$. In other words, MI is a symmetric functional.

In an ideal channel, x is the same as y. This implies that $I(x, y) = H(x) = H(y)$; however, ideal channels are also an extinct species. A real channel usually introduces distortion and noise. From an information-theoretic standpoint, distortion and noise have a direct impact on the entropy of the signal that is distorted or that has noise added to it. The maximum $I(x, y)$ for a certain set of input, output, and channel conditions is the channel capacity, or the number of bits/sec that can be transmitted by the channel. Symbolically, this can be written as

$$C = \max I(x, y). \tag{1.4}$$

Recalling the definitions of the functionals for entropy,

$$H(x) = -\sum_x p(x) \log p(x). \tag{1.5}$$

and conditional entropy

$$H(x|y) = -\sum_x \sum_y p(x, y) \log p(x|y). \tag{1.6}$$

Using the definition of MI from Eq. (1.3) and expanding the above, we get

$$I(x,y) = \sum_x \sum_y p(x,y) \ln \left(\frac{p(x|y)}{p(x)} \right). \tag{1.7}$$

Recall that

$$p(x|y) = \frac{p(x,y)}{p(y)}. \tag{1.8}$$

Substituting this into the previous equation, we get

$$I(x,y) = \sum_x \sum_y p(x,y) \ln \left(\frac{p(x,y)}{p(x)p(y)} \right). \tag{1.9}$$

Of course, $I(x,y)$ is a function of the channel's statistical characteristics [which are captured implicitly by $p(x|y)$], the PDF of the transmitted signal x, and the PDF of the received signal y. For a specified or measured set of statistical characteristics, the channel capacity can be computed using the above equation to find the maximum value for $I(x,y)$.

While the communications channel model for interpreting MI is illustrative and insightful, we have still not introduced a formulation of the restoration problem that can be solved using information theory. Clearly, if the process of image acquisition were shown to be analogous to that of the communications channel, then one can directly apply the above discussion to image restoration. Frieden[9,10] was the first to postulate the image restoration problem as a form of the communication channel problem. Frieden assumed that the channel characteristics are known, and proposed that the best restoration of an image is that for which the input PDF maximizes the MI between the input and output PDFs. Stated in a different way, Frieden proposed that the best solution is that for which the channel capacity is the maximum.

The idea of minimizing MI first appeared in some later work done by Shannon where he examined problems where the output of a channel cannot be uniquely reconstructed. In such problems, an approximate reconstruction can be obtained, and the accuracy of the approximation is measured by means of a fidelity criterion. This work led to the development of the now well-known field of Rate–Distortion theory.

The present authors have developed a general theory that uses the concept of minimizing MI for probabilistic modeling and estimation. These problems will be examined closely later in this book. Before proceeding to the next topic, we would like to review another very important functional in information theory, relative entropy:

$$I(f,p) = \sum_x \sum_y f(x) \ln \left(\frac{f(x)}{p(y)} \right). \tag{1.10}$$

Relative entropy is encountered in the literature under a variety of names. The term cross entropy is due to Good[11] and has been used by Shore[12,13] and Tzannes,[14] among others. Other names include directed divergence[15] and discrimination information. More recently it is being referred to as relative entropy,[16–18] the interpretation being the entropy of one PDF relative to another. Relative entropy has one very important interpretation from a coding theory perspective; it directly measures the penalty for assuming the wrong PDF, i.e., it tells us how many extra bits we use when we develop a code for a signal, assuming the PDF is f when in reality it is p. At the same time, the relative entropy is not symmetric; $I(f, p) \neq I(p, f)$. A generalization of relative entropy is the symmetric J-Divergence of two PDFs, p and q, defined as:

$$J(p, q) = I(p, q) + I(q, p). \tag{1.11}$$

The J-divergence has all of the properties of a distance (or metric) as defined in topology except the triangle inequality property. It is therefore not a distance[15] in the strict sense of the term but is a useful measure of the separation of two distributions. It still preserves the convexity property, and unlike the relative-entropy functional, it is a symmetric functional of the densities.

References

1. R. Gallager, *Information Theory and Reliable Communication*, 1st ed., John Wiley & Sons, New York, NY (1968).

2. F. Ingels, *Information and Coding Theory*, 1st ed., Intext Educational Publishers, New York, NY (1971).

3. C. Shannon, "A mathematical theory of communication," *Bell Systems Technical Journal* **27**, 379–423, 623–656 (1948).

4. E. Jaynes, "Information theory and statistical mechanics," *Physical Review* **106**(4), 620–630 (1957).

5. E. Jaynes, "New engineering applications of information theory," in *Proceedings of the Symposium of Engineering Applications of Random Function Theory and Probability*, Wiley (1963).

6. E. Jaynes, "Prior probabilities," *IEEE Transactions on Systems Science and Cybernetics* **4**, 227–241 (1968).

7. F. Stremler, *Introduction to Communication Systems*, 3rd ed., Prentice-Hall, Upper Saddle River, NJ (1990).

8. S. Haykin, *Communication Systems*, John Wiley & Sons, New York, NY (1983).

9. B. R. Frieden, "Image restoration using a norm of maximum information," *Optical Engineering* **19**, 290–296 (1980).

10. B. R. Frieden, "Maximum-information data processing: application to optical signals," *JOSA* **71**(3), 294–303 (1981).

11. I. Good, "Maximum entropy for hypothesis formulation, especially for multi-dimensional contingency tables," *The Annals of Mathematical Statistics* **34**(3), 911–934 (1963).

12. J. Shore and R. Johnson, "Axiomatic derivation of the principle of maximum entropy and the principle of minimum cross-entropy," *IEEE Transactions on Information Theory* **26**(1), 26–37 (1980).

13. J. Shore and R. Johnson, "Properties of cross-entropy minimization," *IEEE Transactions on Information Theory* **27**(4), 472–482 (1981).

14. N. Tzannes and M. Tzannes, "Transform coded data reconstruction using minimum cross entropy," in *Proceedings of the Information Systems and Sciences Conference*, (1986).

15. S. Kullback, *Information Theory and Statistics*, Dover Publications, Mineola, NY (1997).

16. M. Tzannes and J. Noonan, "On a relation between the principle of minimum relative entropy and maximum likelihood estimation," in *IEEE International Symposium on Circuits and Systems*, **3**, 2132–2135 (1990).

17. J. Shore, "On a relation between maximum likelihood classification and minimum relative-entropy classification," *IEEE Transactions on Information Theory* **30**(6), 851–854 (1984).

18. J. Noonan and M. Tzannes, "An iterative algorithm for entropy maximization and relative entropy minimization," in *Proceedings of IEEE Second Biennial ASSP MiniConference*, 56–58, Boston, MA (1989).

Chapter 2
The Inverse Problem

2.1 Introduction

Given a list of effects, the problem of determining cause has intrigued philosophers, mathematicians and engineers throughout recorded history. Problems of this type are formally referred to as *inverse problems*. Inverse problems pose a particularly difficult challenge: no solution is guaranteed to be unique or stable. The solution is unique only if for some reason *known to the observer* the given list of effects can be due to one and only one cause.

We are concerned here with the inverse problem as it relates to signal and image restoration. In this context of linear time-invariant (LTI) systems, it is common to use the terms *inverse problem* and *deconvolution* interchangeably. The problem here may be stated as that of estimating the true signal given a distorted and noisy version of the true signal.

2.2 Signal Restoration

In general, the goal of signal recovery is to find the best estimate of a signal that has been distorted. Although the mathematics is the same, we would like to distinguish between signal restoration and signal reconstruction. In the first problem, the research is concerned with obtaining a signal that has been distorted by a measuring device whose transfer function is available. Such a problem arises in image processing, wherein the distorting apparatus could be a lens or an image grabber. In the second problem, the scientist is faced with the challenge of reconstructing a signal from a set of its projections, generally corrupted by noise. This problem arises in spectral estimation, tomography, and image compression. In the image-compression problem, a finite subset of projections of the original signal are given, perhaps on the orthonormal cosine basis, and the original signal is desired.

Generally, to go about the problem of signal recovery, a mathematical model of the signal-formation system is needed. Different models are available; simple linear models are easy to work with but do not reflect the real world. More realistic models are complex and may be used at some additional computational cost.[1]

Once a model is specified, a recovery criterion must be selected. Many such criteria exist—ME, minimum-mean-squared error (MMSE), maximum likelihood

(ML), and maximum *a posteriori* (MAP) probability are but a few criteria that have proved useful. Mathematically, the problem of signal recovery is referred to as solving an inverse problem. Generally, an inverse problem will be characterized as being either *well posed* or *ill posed*.[2,3] We clarify these notions below.

Typically, signal restoration/reconstruction belongs to the class of ill-posed problems. That is, we are often concerned with inverting a singular or nearly singular operator. Our goal is to convert a real problem into one which is well posed in the sense that the statement of the problem gives just enough information to determine one unique solution. However, the new reformulation of the problem is often unrealistic due to the assumptions made to choose a particular model or prior information used to tackle the problem. Many attempts have been made to deal with such problems by inventing *ad-hoc* algorithms that imitate this direct mathematical inversion that one would like to carry out. We shall examine these problems in detail.

As stated earlier, signal-restoration techniques seek the best estimate of the true signal given the observed signal and other *a priori* data such as the noise variance, noise PDF, or positivity constraints on the signal itself. The performance of the Gerchberg–Saxton algorithm[4] illustrates the degree of success one can achieve in reconstructing the original signal given only the magnitude of its Fourier transform along with the knowledge that the signal is nonnegative.

In recent years deconvolution has become a science in itself. Extremely sophisticated mathematical techniques have been applied to the solution of this problem. Each of the resulting algorithms has its advantages and pitfalls. Experimentation has confirmed that the success of a given algorithm is intimately related to the characteristics of the data. Thus the need for newer and more general algorithms remains.

The objective of this work is to present and analyze a new generalized formulation for iterative signal restoration. The generalized mapping function (GMF) is presented, and its convergence is studied both in the general formulation and for specific cases. The van Cittert algorithm is a special case of this mapping function. The convergence of the van Cittert algorithm has been discussed by Hill and Ioup[5] and by Jansson.[6] We present a novel and elegant method of obtaining the criteria for convergence of this algorithm. This also serves as a check to establish the validity of the general formulation. Further, we demonstrate that some popular algorithms are special cases of the GMF for most practical purposes. The convergence of these algorithms is analyzed using the structure developed in this work. A few examples are presented, and the direction of our future work is described.

2.3 Well-Posed and Ill-Posed Problems

Many modern experimental devices for investigating physical phenomena and objects of different kinds are complicated. The results of observations are to be processed and interpreted to extract the necessary information about the characteristics of the phenomenon or object to be studied.

Most often, what is measured in a physical experiment is not the desired parameter, represented here by the vector $\mathbf{x}$, but instead a certain effect, $\mathbf{y} = \mathbf{Hx}$. Therefore, the interpretation problem usually reduces to solving an algebraic equation of the form

$$\mathbf{H} \cdot \mathbf{x} = \mathbf{y} \tag{2.1}$$

for the unknown vector $\mathbf{x}$ of length N. Usually, $\mathbf{H}$ represents the apparatus function matrix $(N \times N)$, often termed the impulse transfer function. If the measuring device is linear, then the functional relationship between $\mathbf{x}$ and $\mathbf{y}$ is given by[*]

$$\int_a^b H(t,s)x(s)ds = y(t) \qquad \forall t \in [c,d], \tag{2.2}$$

where the kernel H represents the measuring device and is assumed known. The integral equation is a Fredholm integral of the first kind.

For the following discussion, let us assume that the unknown function $x(s)$ belongs to a metric space[†] F and the known function $y(t)$ to a metric space U. Also, assume the kernel $H(s,t)$ is continuous with respect to t, and that it has a continuous partial derivative $\partial H / \partial t$. Usually, we measure changes in both spaces with the L_2 metric defined by Avriel:[7]

$$\rho_c(\omega_1(t), \omega_2(t)) = \left(\int_c^d [\omega_1(t) - \omega_2(t)]^2 \right)^{0.5} \tag{2.3}$$

in the continuous domain. For the discrete field, we have

$$\rho_D(\omega_1(t), \omega_2(t)) = \left(\frac{1}{N} \sum_{i=1}^N [\omega_1(t) - \omega_2(t)]^2 \right)^{0.5}. \tag{2.4}$$

In the classical sense, solving for x is equivalent to finding the inverse operator H^{-1}, which leads to:

$$\mathbf{x} = \mathbf{H}^{-1}\mathbf{y}. \tag{2.5}$$

Obviously, Eq. (2.1) has solutions for functions $\mathbf{y}$ that lie in the image space $\mathbf{HF}$. Since the right-hand member $y(t)$ is usually obtained experimentally, only an approximation is available and the apparatus function is only known to some given accuracy. Thus, we are faced with the challenging problem of solving for $\mathbf{x}$ when only partial or approximate information is available. Hence, we are dealing with a system:

$$\tilde{\mathbf{H}}\mathbf{x} = \tilde{\mathbf{y}}, \tag{2.6}$$

---

[*]The notation used in this work is described in Appendix D.
[†]metric space: see Appendix C for definition.

which deviates from the initial equation given in Eq. (2.1). Specifically,

$$||\tilde{H} - H|| \leq \delta, \qquad ||\tilde{y} - y|| \leq \tau, \tag{2.7}$$

where the norm is arbitrary and δ, τ are some positive numbers. The question then arises: is the approximate system solvable? Frequently, the operator H is not invertible or its inverse is not continuous (when H is everywhere continuous). Then, the problem on hand is termed "ill posed."

Definition 2.3.1. The problem of determining the solution $\mathbf{x}$ in the space F from the "initial data" $\mathbf{y}$ in the space U is said to be ill posed on the pair of metric spaces (F, U) if at least one of the following three conditions is violated:

1. For every element y in U there exists a solution x in the space F.

2. The solution is unique.

3. The problem is stable in the spaces (F, U).

The property of stability is defined as follows.

Definition 2.3.2. The problem of determining the solution $\mathbf{x} = \mathbf{R}(\mathbf{y})$ in the space F from the initial data $\mathbf{y}$ in U is said to be stable on the spaces (F, U) if, for every positive number ϵ, there exists a positive number $\delta(\epsilon)$ such that the inequality $\rho_U(y_1, y_2) \leq \delta(\epsilon)$ implies $\rho_F(x_1, x_2) \leq \epsilon$ where $x_i = R(y_i)$ with y_i in U and x_i in F for $i = 1, 2$.

Finally, throughout this work we will consider the case when the transformer or operator H is homogeneous. That is, the general integral equation becomes

$$\int_{-\infty}^{\infty} H(t - \tau)x(\tau)d\tau = y(t), \tag{2.8}$$

known as the Fredholm integral equation of the first kind.[8] Any scanning measurement device leads to this form of the integral equation in a noise-free situation.

In the presence of noise, the models presented above are modified to account for the effect of the noise. Also, we will take as a given that the distorting processes are time (or space) invariant. With these changes the model using signal notation is as in Eqs. (2.9) (2.10) below

$$y(t) = \int_{-\infty}^{\infty} h(t - \tau) x(\tau) d\tau + n(t), \tag{2.9}$$

which may be written more compactly as

$$y = h * x + n. \tag{2.10}$$

In Eq. (2.10), $y, x, h,$ and n could be continuous functions of time or their sampled versions. We develop algorithms in this book under the assumption that the noise n may follow any distribution. On the other hand, when verifying these algorithms and the applications they admit, we assume a normal distribution on the noise. We use a minimax argument to justify this assumption. As argued in Ref. 9, this assumption provides the largest lower bound for the variance of any unbiased estimator of x for a general class of linear models. Consequently, assuming noise normality provides the worst-case scenario under which to investigate our estimation and restoration algorithms.

The same model may be written in matrix notation as

$$\mathbf{y} = \mathbf{H} \cdot \mathbf{x} + \mathbf{n}, \tag{2.11}$$

where $\mathbf{y}$ is the sample observation vector, $\mathbf{x}$ is a sample vector representing the true signal, $\mathbf{H}$ is the distorting function matrix, and $\mathbf{n}$ is the sample noise vector. The distortion process is the standard convolution integral. In the matrix model, $\mathbf{H}$ is a circulant Toeplitz matrix and the matrix product in Eq. (2.11) specifies discrete convolution. The noise variance is assumed to be σ^2.

In the case of two-dimensional signals such as images, the models are the same except that the functions are now over two variables. There is a slight modification to the structure of $\mathbf{H}$ and the vectors $\mathbf{x}$, $\mathbf{y}$ in the matrix model; $\mathbf{H}$ is a block Toeplitz matrix, and the vectors $\mathbf{x}$, $\mathbf{y}$ are lexicographically ordered.[10] Lexicographic ordering of image matrices is described in Appendix A.

2.4 Naïve Approaches to Inverse Problems

Our interest is in estimating the process x given the measurement y. The naïve approach is to find the inverse of the distorting function and ignore the effect of noise. In the matrix case this involves computing the inverse of the distorting matrix, i.e.,

$$\mathbf{x} = \mathbf{H}^{-1} \cdot \mathbf{y}. \tag{2.12}$$

This approach runs into difficulties for any but the most trivial cases. The problem lies in determining the inverse matrix. There is, as we have stated before, no guarantee that the inverse exists. In cases where the matrix is noninvertible there is scope for preconditioning it to obtain an inverse, but such approaches may be unstable.

Define h to be the sampled distortion function vector. Let x, y, and n be the true sampled signal vector, measurement vector, and sampled noise vector respectively. Then, instead of using the Toeplitz form Eq. (2.11), the degradation process can be written as

$$y = h * x + n. \tag{2.13}$$

In the frequency domain this is represented as

$$Y = H \cdot X + N. \tag{2.14}$$

Direct inversion implies dividing by H and ignoring the effect of noise to obtain

$$\frac{Y}{H} = X. \tag{2.15}$$

This method assumes that the noise is negligible, and thus by definition is valid only for high signal-to-noise ratio (SNR) environments. Even so, there are many problems associated with this technique. The first and foremost is that H is usually zero or close to zero over a significant portion of the frequency domain. At such points the division by H is meaningless or causes unstable behavior, i.e., bounded changes in the input cause unbounded changes in the output. It also has the effect of significantly amplifying the effect of noise wherever or whenever the true signal has a small magnitude. The filter

$$F = \frac{1}{H} \tag{2.16}$$

is called the Fourier inverse filter.

It is important to note that inverse filters fail to make any use whatsoever of the properties of x itself. See Biemond[11] for an example, where inverse filtering actually increases the distortion in the observed image. When the problem is ill posed, restoration via pseudoinversion suffers from similar problems. Since the resulting solution must have the smallest norm of all possible solutions, it may suffer the most from added noise in the observation.

2.5 Conclusion

The problem of estimating the true states of natural processes is necessarily ill posed, since direct observations of these states are rarely available. Typically, *ground truth* is distorted by the observation process. Direct inversion of the modeled distortion is an intuitive strategy for solving ill-posed problems but, ultimately, it is often far too simplistic an approach in dealing with uncertainty inherent in such problems.

In the next chapter, standard techniques of signal and image restoration are reviewed. For the sake of clarity the discussion is largely focused on image restoration, but the same approaches can be used for one-dimensional signals.

References

1. H. Trussell and M. Civanlar, "The feasible solution in signal restoration," *IEEE Transactions on Acoustics, Speech and Signal Processing* **32**(2), 201–212 (1984).

2. N. Karayiannis and A. Venetsanopoulos, "Regularization theory in image restoration-the stabilizing functional approach," *IEEE Transactions on Acoustics, Speech and Signal Processing* **38**(7) (1990).

3. A. Tikhonov and V. Arsenin, *Solutions of Ill-Posed Problems*, V.H. Winston & Sons, Washington, DC (1977).

4. R. Gerchberg and W. Saxton, "A practical algorithm for the determination of phase from image and diffraction plane pictures," *Optik* **35**, 237 (1992).

5. N. Hill and G. Ioup, "Convergence of the van Cittert iterative method of deconvolution," *JOSA* **66**(5), 487–489 (1976).

6. P. Jansson, *Deconvolution: With Applications In Spectroscopy*, Academic Press, San Diego, CA (1995).

7. M. Avriel, *Nonlinear Programming: Analysis and Methods*, Dover Publications, Mineola, NY (2003).

8. S. Twomey, "On the numerical solution of Fredholm integral equations of the first kind by the inversion of the linear system produced by quadrature," *Journal of the ACM* **10**(1), 101 (1963).

9. P. Stoica and P. Babu, "The gaussian data assumption leads to the largest cramér-rao bound [lecture notes]," *Signal Processing Magazine, IEEE* **28**(3), 132–133 (2011).

10. A. Jain, *Fundamentals of Digital Image Processing*, Prentice-Hall, Upper Saddle River, NJ (1989).

11. J. Biemond, R. Lagendijk, and R. Mersereau, "Iterative methods for image deblurring," *Proceedings of the IEEE* **78**(5), 856–883 (1990).

Chapter 3
Review of Signal-Restoration Techniques

3.1 Introduction

Many researchers have proposed different approaches for overcoming the obstacles to solving the inverse problem. Arguably the most popular classical technique is the Wiener–Kolmogorov filter technique.[1] This technique is implemented in the Fourier domain using the power spectra of the respective images and functions. Modern methods of image restoration are most often based on regularization theory. At the heart of these techniques is the idea of conditioning the solution in a certain manner so as to make it stable.

In the rest of this work m, n will denote spatial domain coordinates, while u, v will denote frequency-domain coordinates. Also, small letters represent functions in the spatial domain whereas their capitalized counterparts represent functions in the frequency domain. Letters with a hat on top, such as $\hat{z}$, emphasize the fact that the symbol represents an estimate of the value or function.

3.2 Wiener–Kolmogorov Filters

We have used the Fourier inverse filter $\mathbf{F_i}(\mathbf{u}, \mathbf{v})$ to obtain an estimate for the process x given the observed process y. The Wiener–Kolmogorov filter $F_w(u, v)$ overcomes the stability problems associated with direct inverse filters. This filter was proposed by Bracewell and Heltron based on the work of Wiener and Kolmogorov.[*] A notable feature of the filter is that it is the MMSE linear filter. The filter uses the ratio of the power spectra of the image and noise to prevent noise amplification. It is given by the following equation

$$\mathcal{F}(u, v) = \frac{H^*(u, v)\, S_{xx}}{|H|^2\, S_{xx} + S_{nn}}, \tag{3.1}$$

where S_{xx} and S_{nn} are the power spectra of the object and noise.

---

[*]In fact, the theory for the discrete case was developed entirely by Kolmogorov, yet the name Wiener filter seems to have become the standard.[2]

This explicit dependence on the availability of the respective spectra is a limitation on the utility of this method. The presence of S_{nn} in the denominator obviously suppresses those frequencies where the noise spectra is dominant. While being a natural course to follow, this has the effect of *smoothing* the resultant estimate since the suppressed frequencies tend to be towards the higher end of the spectrum.

Frieden proposed the so-called sharpness-constrained filter as a means of enhancing (or optionally suppressing) the higher frequencies, thereby improving the sharpness of the estimate. The sharpness-constrained filter is described in Chapter 2 of Ref. 3. The other important filter used for signal restoration is the Kalman filter. See Ref. 4 for details.

3.3 Constrained Least-Squares Restoration

Unlike the inverse filter, which tends to overfit to the data, the constrained least-squares restoration (CLSR) technique is based on the observation that there is no gain in trying to perform better than allowed by the noise interference. So the restoration is constrained to satisfy

$$\| Y(u,v) - H(u,v) \hat{X}(u,v) \| = \| N(u,v) \|, \tag{3.2}$$

where $\| \cdot \|$ denotes the L_2 norm.

Obviously the above equation is just a constraint on the restoration and does not offer a method for obtaining the restored image itself. In CLSR the restoration $\hat{x}$ is chosen such that it minimizes the function $\Omega(\hat{x})$

$$\Omega\left(\hat{X}\right) = \| C(u,v) \hat{X}(u,v) \|, \tag{3.3}$$

subject to the condition that the constraint in Eq. (3.2) is satisfied. Here, $C(u,v)$ is the Fourier transform of the point spread function (PSF) $c(m,n)$ of an operator that measures the non-smoothness of the restoration.[5] An example of such an operator is the gradient. We note here that the constraint presented in this section is of importance to our work since the same constraint is used in deriving the GMF.

3.4 Bayesian Restoration

It is possible to formulate the problem of image restoration as an estimation problem. Using this formulation one may then bring to the problem established techniques from estimation theory such as ML estimation and MAP estimation. Assume that the $p(x)$ and $p(y)$ are the probability distribution functions of the object and its estimate. The MAP estimate is that x belonging to the set of feasible solutions[†] that is most probable given the observation y. Thus, the MAP estimate $\hat{x}_{MAP}$

[†]See Chapter 7 for a definition of the feasible set of solutions.

maximizes

$$p\left(x \mid y\right) \ = \ \frac{p\left(y \mid x\right)\,p\left(x\right)}{p\left(y\right)}, \tag{3.4}$$

and $p\left(x \mid y\right)$ is called the posterior density. This approach is used quite widely in estimation theory. The MAP estimate is the mode of the posterior density.

The maximum likelihood estimate (MLE) $\hat{x}_{ML}$ is that x which is most likely to have caused the observation y. For a detailed description of Bayesian image restoration, see Ref. 6.

3.5 Maximum-Entropy Reconstruction

Jaynes showed that the least-biased estimate of a probability distribution function[‡] given the first few moments of the function is the one that maximizes the entropy of the function. In the discrete case when X is a finite set whose elements are denoted by x_i and their probability of occurrences by p_i, the entropy is defined as

$$E \ = \ \sum_{i=0}^{i=n} p_i \cdot \frac{1}{ln\left(p_i\right)}, \tag{3.5}$$

and if X has infinite cardinality, then

$$E \ = \ \int_{x=-\infty}^{x=\infty} p_x\left(x\right) \cdot \frac{1}{ln\left[p_x\left(x\right)\right]}, \tag{3.6}$$

where $p_x\left(x\right)$ is the probability distribution function of x.

Frieden and Burg extended this technique to the restoration of images. At present there are different formulations of the ME approach to image restoration. We shall present one form in Section 6.2 to give an indication of how the method is implemented.

The ME approach simply selects the solution, consistent with the constraints, that has the most entropy. Of course, it is necessary to have available a feasible set of solutions before the one with the most entropy may be chosen. For more details refer to Frieden,[7] Wernecke,[8] or Weir.[9]

Since iterative schemes of restoration are central to this work, we will present the relevant material in detail. Iterative schemes have been applied to the problem of recovering a signal from its distorted version for quite a few years. In fact, we are not sure as to the first application of this approach to signal restoration. In Chapter 2, we presented the direct inverse filter solution and the reasons for its inadequacy. Hence, alternative approaches are important. The techniques presented in Chapter 3 for overcoming these inadequacies are limited in their ability to incorporate *a priori* information about the nature of the true signal or image. Iterative methods offer the

---

[‡]See Appendix A for a discussion of the general problem of estimating a function by the projection method.

possibility of taking into account available information such as the nonnegative or the bandlimited nature of the solution. Our development of the topic follows the approach in Biemond[5] and Schafer.[10] For notational convenience we will use the matrix formulation given in Eq. (2.11).

Additional notation is introduced for the object space containing $\mathbf{x}$ and the image space containing $\mathbf{y}$. The former is denoted by $\mathcal{S}$ and the latter by $\mathcal{V}$.

3.6 The Iterative Technique

The basic iteration equation is

$$\mathbf{x}_{k+1} = T\mathbf{x}_k, \tag{3.7}$$

where T is an operator that is obtained by using our distortion model and any other *a priori* information that is available. Such an equation results from determining a solution to the mapping

$$\mathbf{x} = T\mathbf{x}, \tag{3.8}$$

by the method of successive substitutions. Let us for the moment neglect the effect of noise on our universe[§] and let us assume that all *a priori* information can be represented by a constraint operator, say $\mathbf{C}$. Then

$$\mathbf{x} = \mathbf{C}\mathbf{x} \tag{3.9}$$

is satisfied if and only if $\mathbf{x}$ has the properties specified in the constraints. Commonly specified properties are

- The positivity of image intensities or spectral amplitudes:

$$\mathbf{x}[i] \geq 0 \ \forall i \in \{1, 2, \ldots, n\}.$$

- The bandlimited nature of most signals–especially in the case of discrete sampled signals.

The operator $\mathbf{C}$ effects no change on those signals that are consistent with the *a priori* information, whereas the signals that are not similarly endowed are modified so that they conform to the constraints. We may then write

$$\mathbf{y} = \mathbf{H} \cdot \mathbf{C}\mathbf{x}, \tag{3.10}$$

if $\hat{\mathbf{x}}$ is an estimate of $\mathbf{x}$, then the error $\mathbf{e_y}$ signal in the space $\mathcal{V}$ may be written as

$$\mathbf{e_y} = \mathbf{y} - \mathbf{H} \cdot \mathbf{C}\hat{\mathbf{x}} \tag{3.11}$$

[§]Early iterative methods were all *ad hoc*. As such, the consideration of noise was precluded. In the context of these iterative equations, it is common practice in the literature to ignore the effect of noise.

Assuming for the moment that the error signal e_y in $\mathcal{V}$ is mapped to a e_x in $\mathcal{S}$ by a simple constant gain factor, we may write

$$\mathbf{x} = \mathbf{Cx} + \lambda(\mathbf{y} - \mathbf{H} \cdot \mathbf{Cx}), \qquad (3.12)$$

which when rearranged yields

$$\mathbf{x} = \mathbf{Tx} = \lambda\mathbf{y} + (\mathbf{I} - \lambda\mathbf{H}) \cdot \mathbf{Cx}. \qquad (3.13)$$

This is the basic iterative equation for recursively solving for $\mathbf{x}$ given the data $\mathbf{y}$. The recursive nature makes it necessary to consider the convergence of the scheme and the uniqueness of the solution.

3.6.1 Van Cittert iteration

The Van Cittert[11] restoration technique is the simplest and probably the earliest example of applying iterative techniques to deconvolution. The simple recursion governing the method is

$$\hat{\mathbf{x}}_{k+1} = \hat{\mathbf{x}}_k + \lambda(\mathbf{y} - \mathbf{H} \cdot \hat{\mathbf{x}}_k). \qquad (3.14)$$

This method has been extensively studied due to both its simplicity and the pleasing nature of the results in the context of iterative restoration. We shall see that this technique is a special case of the GMF.

It is important to note here that the method of successive substitutions is not the only possible approach for solving equations of the form of Eq. (3.8). In fact, Eq. (3.14) may be rewritten after some straightforward manipulation[5] as

$$\hat{\mathbf{x}}_{k+1} = \beta(\mathbf{I} - \mathbf{R})^{-1}\left(\mathbf{I} - \mathbf{R}^{k+1}\right) \cdot \mathbf{y}, \qquad (3.15)$$

where $\mathbf{R} = (\mathbf{I} - \beta\mathbf{H})$, provided $(\mathbf{I} - \mathbf{R})$ is invertible. A sufficient condition for convergence is

$$\lim_{k \to \infty} \mathbf{R}^{k+1}\mathbf{g} = 0. \qquad (3.16)$$

Assuming this is true, the following equality holds

$$\hat{\mathbf{x}}_\infty = \lim_{k \to \infty} \hat{\mathbf{x}}_k = \beta(\mathbf{I} - \mathbf{R})^{-1} \cdot \mathbf{g} = \mathbf{H}^{-1} \cdot \mathbf{g}, \qquad (3.17)$$

which is precisely the inverse filter. It is then reasonable to ask the following question: what makes the Van Cittert method any better than the inverse filter? The answer to this lies in the observation that only if the iteration is allowed to converge is the Van Cittert estimate indistinct from the inverse filter solution. But the iterative method offers the possibility of stopping the iterations at any convenient step, and this is exactly what is done. Further, instead of computing the inverse filter we simply blur the estimate repeatedly to generate an error signal. This illustrates the power of the iterative method in conditioning the solution to be stable. We shall come back to this simple technique at a later stage.

3.7 Conclusion

The extent of the literature on the subject of inverse problems is important to note. Therefore, our discussion of the solution methods to the deconvolution problem was intended to emphasize the diversity of the methods rather than to give an exhaustive list of techniques. Very successful but complicated methods were omitted for the sake of clarity.

In general, as should be apparent from the review of the traditional methods given above, solving and regularizing different kinds of inverse problems essentially involves a compromise between fidelity to the data and fidelity to the available prior information about the sought solution. This compromise can be qualified through the use of an optimality criterion. Tikhonov[12] was the first one to set up the mathematical foundations to the regularization concept of ill-posed problems. The mathematical formalism is known as regularization theory. The choice of the optimality criterion is known as the stabilizing-function approach. A few popular functionals are the Euclidean and weighted quadratic distances, the Kullback–Leibler (KL) distance, and local energy functions. The successful use of these optimality criteria encouraged the present authors to investigate further the choice of the distances. Application of the stabilizing functional approach to the class of generalized entropic distances, rooted in the measure theory of mathematics and the MI function of information theory, produced new classes of algorithms. In later chapters, we formally present the distance measures upon which these classes of algorithms are based.

References

1. A. Jain, *Fundamentals of Digital Image Processing*, Prentice-Hall, Upper Saddle River, NJ (1989).

2. H. Stark and J. Woods, *Probability, random processes, and estimation theory for engineers*, Prentice-Hall, Upper Saddle River, NJ (1986). *Ch 2 - Five Lin*

3. P. Jansson, *Deconvolution: With Applications In Spectroscopy*, Academic Press, San Diego, CA (1995).

4. J. Woods and V. Ingle, "Kalman filtering in two dimensions: Further results," *IEEE Transactions on Acoustics, Speech and Signal Processing* **29**(2), 188–197 (1981).

5. J. Biemond, R. Lagendijk, and R. Mersereau, "Iterative methods for image deblurring," *Proceedings of the IEEE* **78**(5), 856–883 (1990).

6. B. Hunt, "Bayesian methods in nonlinear digital image restoration," *IEEE Transactions on Computers* (1977).

7. B. Frieden, "Restoring with maximum likelihood and maximum entropy," *JOSA* **62**, 511–518 (1972).

8. S. Wernecke and L. D'Addario, "Maximum entropy image reconstruction," *IEEE Transactions on Computers* , 351–364 (1977).

9. N. Weir, *MEM: Maximum Entropy Method Image Restoration System*, Caltech, Pasadena, CA (1990).

10. R. Schafer, R. Mersereau, and M. Richards, "Constrained iterative restoration algorithms," *Proceedings of the IEEE* **69**(4), 432–450 (1981).

11. P. Van Cittert, "Zum Einfluß der Spaltbreite auf die Intensitätsverteilung in Spektrallinien," *Zeitschrift für Physik A Hadrons and Nuclei* **65**(7), 547–563 (1930).

12. A. Tikhonov and V. Arsenin, *Solutions of Ill-Posed Problems*, V.H. Winston & Sons, Washington, DC (1977).

Part II

Density Estimation: Maximum Entropy

Chapter 4

Maximum-Entropy Density Estimation

4.1 Introduction

An *uncertain* system yields random outputs even for deterministic inputs. Consequently, design and analysis of such systems require faithful models of this uncertainty. In a statistical framework, these models are typically specified by probability distributions, which are often unknown *a priori* and must therefore be estimated from available data or information. Unfortunately, the observed uncertainty may not be uniquely specified by a single distribution. In this case, unique densities are obtained by constraining candidate densities to satisfy additional optimality criteria. Motivated by the principles of information theory, we now detail how *entropy* may be used as one such constraint.

4.2 Density Estimation

Prior information concerning an uncertain system may comprise of observed data or of a moment sequence, i.e., the average values of a sequence of known statistics. For instance, information about system power (second moment) is often available when modeling communications systems. When available information consists of only observed data, the density model is estimated via parametric or nonparametric methods. In the parametric setting, a functional form of the density is assumed (e.g., Gaussian, Laplacian, etc.) and the parameters specifying the model are estimated from the data. Often, this approach requires a *post-hoc* validation of the assumed model.

Typically, nonparametric estimation methods are more robust; no assumption is made regarding the functional form of the density. One commonly used nonparametric density estimation method is the kernel method,[1] wherein a kernel is applied to smooth an ordinary histogram estimate. Some kernel estimators admit adaptive updates of density estimates[2] as new observational data are made available. However, as these estimates do not satisfy any information-theoretic optimality criterion, we will not discuss them in this monograph.

4.2.1 Information-driven estimation

When available information consists only of a sequence of moments, the problem becomes one of recovering the unknown density from its known moment sequence, often called the *moment problem*. This problem was well known and studied extensively even before Shannon introduced the concept of entropy into engineering.

The moment problem does not always have a unique solution; even an infinite sequence $\{\mu_k\}_{k=1}^{\infty}$ of moments may not specify a unique density. To wit, all elements of the collection of densities $\{p_a(x)\}$ with

$$p_a(x) = \frac{1}{\sqrt{2\pi}} x^{-\ln x/2 - 1}(1 + a\sin(2\pi \ln x)) \quad \forall a \in [-1, 1] \qquad (4.1)$$

have the same moment sequences, even though each are distinct. Several sufficient conditions[3] *do* exist for an infinite sequence of moments to uniquely define a density. One such condition guarantees uniqueness when the true density has finite support. Another sufficient condition requires a moment sequence of monomial statistics $\{x_k\}$ to be completely monotonic:[4]

$$\Delta_k \mu_n \triangleq \sum_{m=0}^{k} (-1)^m \binom{k}{m} \mu_{n+m} > 0 \quad n, k \in \mathbb{Z}^+. \qquad (4.2)$$

Unfortunately, the preceding sufficient conditions are rarely satisfied in practice; i.e., the moment problem is typically *ill posed*. Consequently, candidate solutions must be selected according to their fitness under some additional optimality criterion. Each of these additional criteria yields a distinct density under certain regularity conditions satisfied by the moment sequence. In the next section, we consider several examples of such criteria, including ME, minimum divergence distance, and minimum description length (MDL).

4.3 Maximum-Entropy Estimation

Since their acceptance as viable principles for statistical inference, ME and minimum divergence have been used extensively to solve the moment problem for finite sequences. What follows is a brief description of these notions and how they are related.[5]

From the definition of entropy and relative entropy, or *divergence distance*, given in Chapter 1, it is evident that entropy is the negative divergence distance between some PDF $p(x)$ and the uniform distribution. Consequently, maximizing entropy minimizes the divergence distance from the uniform. As methods based on both principles involve optimizing the same objective function, and since, for uniform priors, the maximum-entropy principle and the minimum cross-entropy principle provide the same estimate of an unknown density, we focus here only on the ME approach.

Given some information, in the form of moments of a known sequence of statistics, the ME estimation approach advocates choosing as an estimate the density that is closest, in the sense of relative entropy, to the uniform. Intuitively, the estimate deviates from the uniform, or most uninformative distribution,* only as much as the given information will allow. To detail the construction of the estimate, first assume the moments $\mu_k = E_f(r_k(x))$ are available for an arbitrary sequence $\{r_k(x)\}$ of s statistics, which need not be linearly independent. Among all densities $f(x)$ defined on a support Ω that satisfy the s moment constraints

$$\int_\Omega f(x)dx = 1 \qquad E_f r_k(x) = \int_\Omega r_k(x)f(x)dx = \tau_k \qquad 0 \leq k \leq s-1, \quad (4.3)$$

the one that maximizes the entropy

$$H(f) = -E_f \log f(x) = -\int_\Omega f(x) \log f(x)dx \qquad (4.4)$$

is known as the ME density. The estimate [6] may be given as

$$p_m(x, \boldsymbol{\alpha}) = e^{\langle \boldsymbol{\alpha}, r(x) \rangle - \psi(\boldsymbol{\alpha})}, \qquad (4.5)$$

where ψ is some normalizing function of the vector of *natural parameters* $\boldsymbol{\alpha}$.

In general, the representation of Eq. (4.5) may belong to a high-order *exponential family*. To reduce the order, we may orthogonalize the statistic sequence. To wit, there exists a sequence of some $m(\leq s)$ linearly independent statistics, $\{T_k(x)\}$, and a $s \times m$ scalar matrix $\mathbf{M}$ such that $r(x) = \mathbf{MT}(x)$, and $\boldsymbol{\mu} = \mathbf{M}\boldsymbol{\tau}$, or $\boldsymbol{\tau} = (\mathbf{M}^T\mathbf{M})^{-1}\mathbf{M}^T\boldsymbol{\mu}$, where $r(x)$ and $\boldsymbol{\mu}$ are s-dimensional vectors defined by $\{r_k(x)\}$ and $\{\mu_k\}$, respectively. Similarly, $\mathbf{T}(x)$ and $\boldsymbol{\tau}$ are m-dimensional vectors defined by $\{T_k(x)\}$ and $\{\tau_k\}$. The *minimal representation* of the ME estimate is then given as

$$p_m(x, \boldsymbol{\beta}) = e^{\langle \boldsymbol{\beta}, \boldsymbol{T}(x) \rangle - \psi(\boldsymbol{\beta})}, \qquad (4.6)$$

with $\boldsymbol{\beta}$ denoting the new vector of natural parameters found via a change of frame of $\boldsymbol{\alpha}$.

4.4 Estimating Channel Tap Densities Using Maximum Entropy

The following example[7] illustrates an application of ME methods to communications systems design problems. Assume the standard multiple-input multiple-output (MIMO) received signal model

$$\mathbf{y} = \sqrt{\frac{\rho}{\mathbf{n_t}}}\mathbf{Hx} + \mathbf{n}. \qquad (4.7)$$

---

*This is to be distinguished from the Bayesian notion of least informative priors.

Our task is to estimate the channel when the only prior knowledge available is that the channel has some energy E. To restate the problem, we must estimate $P(\mathbf{H})$ under the condition

$$\frac{1}{n_t n_r} E \left(\sum_i \sum_j |h_{ij}|^2 \right) = E. \tag{4.8}$$

Constraining $P(\mathbf{H})$ to Eq. (4.8) and to the set of all PDFs yields the following cost function for $P(\mathbf{H})$:

$$J(P) = - \int P(\mathbf{H}) \ln P(\mathbf{H}) d\mathbf{H} + \lambda_1 \left[1 - \int \mathbf{P(H)dH} \right]$$

$$+ \lambda_2 \left[\frac{1}{n_t n_r} E_{P(\mathbf{H})} \left(\sum_i \sum_j |h_{ij}|^2 \right) - E \right]. \tag{4.9}$$

Differentiating gives

$$\frac{dJ}{dP} = -1 - \ln P(\mathbf{H}) - \lambda_1 - \lambda_2 \sum_i \sum_j |\mathbf{h_{ij}}|^2 = 0, \tag{4.10}$$

which yields

$$P(\mathbf{H}) = e^{-\lambda_1 - \lambda_2 \sum_i \sum_j |\mathbf{h_{ij}}|^2} = \prod_i \prod_j e^{-\lambda_2 |\mathbf{h_{ij}}|^2 - \frac{\lambda_1 + 1}{n_r n_t}}. \tag{4.11}$$

Thus, the example validates the use of the Gaussian independent and identically distributed (IID) channel model when only the channel energy is known.

In practical communications system design problems, knowledge of the stochastic channel model is essential for creating proper source sequences. Consider that the ergodic capacity of a MIMO channel, $\mathbf{H}$, is given by

$$C' = \max_{\mathbf{R}} E(C(\mathbf{R})) \quad \text{where} \quad \mathbf{C(R)} = \log_2 \left| \mathbf{I} + \frac{\rho}{n_t} \mathbf{HRH^H} \right|, \tag{4.12}$$

and $\mathbf{R} = \mathbf{E(xx^H)}$ is the covariance of the stacked-source sequence. It may be shown[8] that for a Gaussian IID channel model, the choice of $\mathbf{R} = \mathbf{I}$ maximizes the ergodic capacity equation given in Eq. (4.12). However, in cases where there is additional information about the channel available, uncorrelated source sequences are no longer guaranteed to match, or even approximate, the ergodic capacity and Eq. (4.12) must be calculated explicitly.

4.5 Conclusion

We have defined the *moment problem* as one in which an estimate of a probability density must be reconstructed from a given moment sequence encoding known information. The ME approach chooses as an estimate that density which is closest to the uniform distribution and is still consistent with the given information. In general, the given information may not yield compact representations of the estimates; orthogonalization of the statistic sequence enables such representations. In the next chapter, we will characterize the uncertainty of such estimators.

References

1. B. Silverman, *Density Estimation For Statistics and Data Analysis*, Chapman & Hall/CRC, Boca Raton, FL (1998).

2. E. Wegman and H. Davies, "Remarks on some recursive estimators of a probability density," *The Annals of Statistics* **7**(2), 316–327 (1979).

3. C. Rao, *Linear Statistical Inference and Its Applications*, John Wiley & Sons, New York, NY (1973).

4. L. Mead and N. Papanicolaou, "Maximum entropy in the problem of moments," *Journal of Mathematical Physics* **25**(8), 2404–2417 (1984).

5. J. Kapur, *Maximum-Entropy Models in Science and Engineering*, John Wiley & Sons, New York, NY (1989).

6. T. Cover and J. Thomas, *Elements of Information Theory*, John Wiley & Sons, New York, NY (2006).

7. M. Debbah and R. Müller, "MIMO channel modeling and the principle of maximum entropy," *IEEE Transactions on Information Theory* **51**(5), 1667 (2005).

8. I. Telatar and D. Tse, "Capacity and mutual information of wideband multipath fading channels," *IEEE Transactions on Information Theory* **46**(4) (2000).

Chapter 5

Error Bounds for Maximum-Entropy Estimates

5.1 Introduction

In a parametric, or model-based, estimation approach, deriving error bounds on density estimates would be straightforward; the uncertainty of the parameter estimates would fully determine the uncertainty in the overall density estimate. However, in the absence of a parametric model, there is no well-defined approach for finding the goodness of a density estimator. Consequently, we now detail the construction of error bounds and confidence intervals for ME density estimates.

Key to our approach is the representation of densities by finite-order *exponential families*. Using exponential families as probability models of uncertain systems offers many advantages. Often, if two independent random variables have densities belonging to the same exponential family, their joint density will also be a member of this family. As such, exponential family models are easily updated as new information becomes available. Moreover, if $f(x)$ is infinitely differentiable, then, modulo certain pathological cases, it may be approximated with arbitrary precision by some density from an exponential family. In this light, the ME method is highly desirable; its estimates are always members of exponential families.

The linearly independent statistics sequence used to define the exponential family may include the sequence of monomials $\{x^k\}$ and the sequences of orthogonal polynomials. We assume the unknown density is from a canonical exponential family of an unknown but finite order m, and only n moments on n linearly independent statistics $\{T_k(x)\}$ are known. We consider two cases: $m \leq n$ and $m > n$. In the first case, we have an overdetermined model. In the second underdetermined case, even when m is known, there may still be an infinite number of densities from an m-parameter exponential family having the same set of prescribed n moments.

Since $f(x)$ is unknown, providing a bound for the modeling error stands as an open question. To wit, approximating densities with high-order exponential families complicates analysis and leads to confusion regarding outliers. On the other hand, approximating with an inadequate number of moments leads to wrong representations and unacceptably large error in the analysis of uncertain systems. Thus, knowing a bound for the approximation error before adopting a probability model for further analysis of the uncertain system is vital. Using the theory of *partial Legendre transforms* and properties of the exponential families, we show that there exists an upper bound for such errors when m is known.

Our contributions are as follows. For $m \leq n$, we prove that ME estimation will determine a minimal representation of the unknown density from an exponential family of order m exactly. We then establish error bounds on estimates produced when the true density comes from an exponential family whose order exceeds the number of available moments. Finally, when moments are estimated from a large sample of observations drawn from the unknown distribution, we use asymptotic properties of the sample divergence distance, or KL distance,[1] to find a confidence set for the unknown PDF.

5.2 Preliminaries

Legendre transforms have been used extensively in mathematical physics; basic information can be found in Arnold,[2] Courant and Hilbert,[3] and Barndorff-Nielsen.[4] A review of exponential families may be found in Barndorff-Nielsen[4] and essential information regarding them in many text books on statistics including Bickel and Doksum.[5] Here, we briefly note basic concepts and some relevant results.

5.2.1 Legendre transforms

The *conjugate* of a function $\psi(\mathbf{x}) : \mathbb{R}^k \to \mathbb{R}$ is the function $\phi(\mathbf{u}) : \mathbb{R}^k \to \mathbb{R}$ is given by

$$\phi(\mathbf{u}) = \sup_{\mathbf{x} \in \mathbb{R}^k} \langle \mathbf{x}, \mathbf{u} \rangle - \psi(\mathbf{x}), \tag{5.1}$$

$$\phi(\mathbf{u}) = \langle \mathbf{x_0}, \mathbf{u} \rangle - \psi(\mathbf{x_0}) \text{ if and only if } \mathbf{u} = \nabla \psi(\mathbf{x_0}), \tag{5.2}$$

i.e., if $\phi(\mathbf{u})$ attains its supremum at a point $\mathbf{x_0}$, then $\mathbf{u} = \nabla \psi(\mathbf{x_0})$. Conversely, if $\mathbf{x_0}$ solves the gradient equation $\mathbf{u} = \nabla \psi(\mathbf{x})$ for a given $\mathbf{u}$, then $\phi(\mathbf{u})$ attains its supremum at the point $\mathbf{x_0}$.

Definition 5.2.1 (Legendre transform). Assume $\psi(\mathbf{x}) : \mathbb{R}^k \to \mathbb{R}$ is some convex function. Then, we may give the *Legendre transform* of $\psi(\mathbf{x})$ as

$$\phi(\mathbf{u}) \triangleq \langle \mathbf{x}, \mathbf{u} \rangle - \psi(\mathbf{x}) \text{ with } \mathbf{u} = \nabla \psi(\mathbf{x}), \tag{5.3}$$

where $\nabla(\cdot)$ is the gradient operator. For each $\mathbf{u}$, $\mathbf{x}(\mathbf{u})$ is a solution of the equation $\mathbf{u} = \nabla \psi(\mathbf{x})$. The function $\phi(\mathbf{u})$ is a convex function in $\mathbf{u}$ and by the *inverse Legendre transform* we have

$$\psi(\mathbf{x}) = \langle \mathbf{x}, \mathbf{u} \rangle - \phi(\mathbf{u}) \tag{5.4}$$

where, for each $\mathbf{x}$, $\mathbf{u}$ solves the gradient equation $\mathbf{x} = \nabla_\mathbf{u} \phi(\mathbf{u})$.

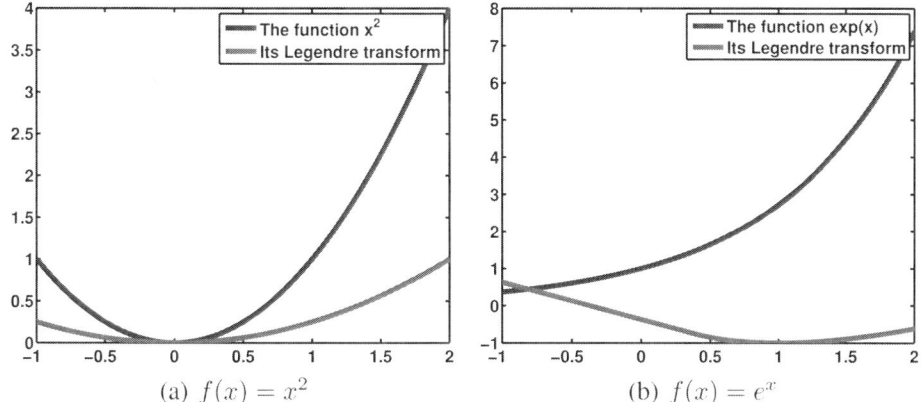

(a) $f(x) = x^2$ (b) $f(x) = e^x$

Figure 5.1 Two examples of Legendre transforms.

Any $\mathbf{x}$ and $\mathbf{u}$ satisfying the relation $\mathbf{u} = \nabla\psi(\mathbf{x})$ or $\mathbf{x} = \nabla\phi(\mathbf{u})$ will also satisfy the relation:

$$\psi(\mathbf{x}) + \phi(\mathbf{u}) = \langle \mathbf{x}, \mathbf{u} \rangle. \tag{5.5}$$

For any other $(\mathbf{x}, \mathbf{u})$, Fenchel's inequality holds:

$$\psi(\mathbf{x}) + \phi(\mathbf{u}) > \langle \mathbf{x}, \mathbf{u} \rangle. \tag{5.6}$$

Provided that the original function is strictly convex, we may interpret the Legendre transform as a mapping from this function to a family of its tangents expressed in functional form. Figs. 5.1(a) and 5.1(b) show two examples of applying the transform to convex functions.

Definition 5.2.2 (partial Legendre transform). Given the following convex function $\psi(\mathbf{x}, \mathbf{y}) : \mathbb{R}^k \times \mathbb{R}^l \to \mathbb{R}$, the *partial Legendre transform* of $\psi(\mathbf{x}, \mathbf{y})$ with respect to $\mathbf{x}$ alone is defined as

$$\phi(\mathbf{u}, \mathbf{y}) \triangleq \sup_{\mathbf{x} \in \mathbb{R}^k} \langle \mathbf{x}, \mathbf{u} \rangle - \psi(\mathbf{x}, \mathbf{y}) = \langle \mathbf{x_0}(\mathbf{u}, \mathbf{y}), \mathbf{u} \rangle - \psi(\mathbf{x_0}, \mathbf{y}),$$

where, for each $(\mathbf{u}, \mathbf{y})$, $\mathbf{x_0}(\mathbf{u}, \mathbf{y})$ is the solution of $\mathbf{u} = \nabla_{\mathbf{x}}\psi(\mathbf{x}, \mathbf{y})$, $\nabla_{\mathbf{x}}(\cdot)$ being the gradient with respect to $\mathbf{x}$ alone.

The partial Legendre transform $\phi(\mathbf{u}, \mathbf{y})$ is convex in $\mathbf{u}$, but a concave function of $\mathbf{y}$; an example of this is illustrated in Fig. 5.2. As such, for each $\mathbf{u}$, $\phi(\mathbf{u}, \mathbf{y})$ is maximized at some $\mathbf{y_0}(\mathbf{u})$, and

$$\phi(\mathbf{u}, \mathbf{y_0}) = \langle \mathbf{x_0}(\mathbf{u}, \mathbf{y_0}), \mathbf{u} \rangle - \psi(\mathbf{x_0}, \mathbf{y_0}) > \phi(\mathbf{u}, \mathbf{y}) = \langle \mathbf{x_0}(\mathbf{u}, \mathbf{y}), \mathbf{u} \rangle - \psi(\mathbf{x_0}, \mathbf{y})$$

for any other $\mathbf{y}$. Also, for any other $\mathbf{x}$, by Fenchel's inequality, we have

$$\phi(\mathbf{u}, \mathbf{y}) > \langle \mathbf{x}, \mathbf{u} \rangle - \psi(\mathbf{x}, \mathbf{y}). \tag{5.7}$$

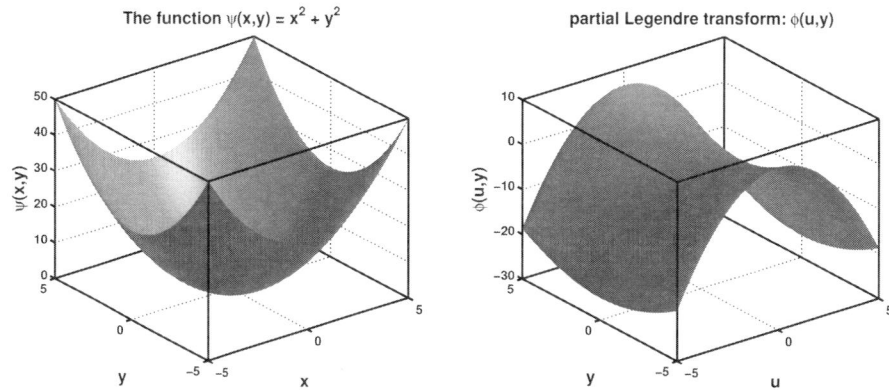

Figure 5.2 An example application of the partial Legendre transform.

The Legendre transform of $\psi(x, y)$ may be written as

$$\phi(\mathbf{u}, \mathbf{v}) = \langle \mathbf{x}(\mathbf{u}, \mathbf{v}), \mathbf{u} \rangle + \langle \mathbf{y}(\mathbf{u}, \mathbf{v}), \mathbf{v} \rangle - \psi(\mathbf{x}, \mathbf{y}),$$

where, for any $(\mathbf{u}, \mathbf{v})$, $\mathbf{x}(\mathbf{u}, \mathbf{v})$ and $\mathbf{y}(\mathbf{u}, \mathbf{v})$ satisfy the gradient equation

$$[\mathbf{u}^{\mathbf{T}}, \mathbf{v}^{\mathbf{T}}]^{\mathbf{T}} = \nabla \psi(\mathbf{x}, \mathbf{y}).$$

Finally, when $\mathbf{u_m} = \nabla_{\mathbf{x}} \psi(\mathbf{x_0}, \mathbf{y_0})$ and $\mathbf{v_m} = \nabla_{\mathbf{y}} \psi(\mathbf{x_0}, \mathbf{y_0})$, we have

$$\phi(\mathbf{u_m}, \mathbf{v_m}) \geq \langle \mathbf{x_0}, \mathbf{u_m} \rangle + \langle \mathbf{y_0}, \mathbf{v_m} \rangle - \psi(\mathbf{x_0}, \mathbf{y_0})$$
$$= \sup_{(\mathbf{x}, \mathbf{y})} [\langle \mathbf{x}, \mathbf{u_m} \rangle + \langle \mathbf{y}, \mathbf{v_m} \rangle - \psi(\mathbf{x}, \mathbf{y})].$$

The pair $(\mathbf{x_0}, \mathbf{y_0})$ that maximizes $\phi(\mathbf{u}, \mathbf{y})$ also maximizes $\phi(\mathbf{u}, \mathbf{v})$.

5.2.2 Exponential families

An exponential family $\mathscr{P} = \{p(x, \boldsymbol{\alpha})\}$ is a collection of density functions admitting an exponential representation

$$p(x, \boldsymbol{\alpha}) = \exp[\langle \boldsymbol{\alpha}, \mathbf{T}(\mathbf{x}) \rangle - \psi(\boldsymbol{\alpha})]$$

with respect to a measure μ on the sample space X for some d. We take this measure to be the Lebesgue measure. All components of the vector of statistics, $\mathbf{T}$, take values in X. The natural parameter space A consists of all $\boldsymbol{\alpha}$ for which the normalizing constant $\psi(\boldsymbol{\alpha})$ is finite:

$$\exp[\psi(\boldsymbol{\alpha})] = \int \exp[\langle \boldsymbol{\alpha}, \mathbf{T}(\mathbf{x}) \rangle] \mathbf{dx} < \infty.$$

The normalizing constant $\psi(\boldsymbol{\alpha})$, sometimes referred to as the log-partition function, uniquely defines a density from an exponential family. A family of exponential densities is called *regular* if its natural parameter space is open with nonempty interior, i.e., $A = A^\circ \neq \emptyset$, where A° denotes the interior of A. All exponential families under consideration here are assumed to be regular. We additionally assume that all generating statistics are linearly independent, this being a necessary condition for the identifiability of the natural parameters.

The proofs of the following facts about exponential families can be found in Refs. 4 or 5:

1. The set of natural parameters A is a convex set.

2. $\psi : A \to \mathbb{R}$ is convex on A.

3. For any $\boldsymbol{\alpha} \in A^\circ$

   - $E_{\boldsymbol{\alpha}}(\mathbf{T}(\mathbf{x})) = \nabla_{\boldsymbol{\alpha}}\psi(\boldsymbol{\alpha})$,
   - $Cov(\mathbf{T}(\mathbf{x})) = \nabla_{\boldsymbol{\alpha}}^2 \psi(\boldsymbol{\alpha})$,
   - $\nabla_{\boldsymbol{\alpha}}^i \psi(\boldsymbol{\alpha}) =$ the ith order cumulant of $\mathbf{T}$.

4. $\log(p(x, \boldsymbol{\alpha}))$ is strictly concave, hence $p(x, \boldsymbol{\alpha})$ along with all of its marginals are strongly unimodal.

5. If X and Y are two independent random variables with respective PDFs $p_1(x, \boldsymbol{\alpha})$ and $p_2(x, \boldsymbol{\beta})$, then their joint density is $p(x, \boldsymbol{\alpha} + \boldsymbol{\beta})$, provided $(\boldsymbol{\alpha} + \boldsymbol{\beta})$ is in the natural parameter space of the m-parameter exponential family.

6. If $p(x, \boldsymbol{\alpha}) = \exp(\langle \boldsymbol{\alpha}, \mathbf{T}(\mathbf{x}) \rangle - \psi_{\mathbf{m}}(\boldsymbol{\alpha}))$, then $\boldsymbol{\alpha}$ and $\boldsymbol{\tau} = E(\mathbf{T}(\mathbf{x}))$ satisfy the Legendre transform relations:

$$\phi(\boldsymbol{\tau}) = \langle \boldsymbol{\alpha}, \boldsymbol{\tau} \rangle - \psi_m(\boldsymbol{\alpha}) \quad \text{and} \quad \psi_m(\boldsymbol{\alpha}) = \langle \boldsymbol{\alpha}, \boldsymbol{\tau} \rangle - \phi(\boldsymbol{\tau}).$$

   Consequently, $\boldsymbol{\tau} = \nabla_{\boldsymbol{\alpha}}\psi_m(\boldsymbol{\alpha})$ and $\boldsymbol{\alpha} = \nabla_{\boldsymbol{\tau}}\phi(\boldsymbol{\tau})$.

7. $\nabla_{\boldsymbol{\alpha}}\psi_m$ and $\nabla_{\boldsymbol{\tau}}\phi$ are inverse transforms of each other; i.e., $\nabla_{\boldsymbol{\alpha}}\psi_m$ is bijective.

5.2.3 Dual representations of exponential family members

The above reflexive Legendre transform relationship implies that each density from an exponential family admits a dual representation involving either the natural parameters $\boldsymbol{\alpha}$ or the mean value or moment parameters $\boldsymbol{\tau} = E_{\boldsymbol{\alpha}}(\mathbf{T}(\mathbf{x}))$; one parameter set uniquely determines the other. Amari[6] refers to the Legendre transform pair $\psi_m(\boldsymbol{\alpha})$ and $\phi(\boldsymbol{\tau})$ as the convex potential functions associated with the natural and moment-parametric representations $p(x, \boldsymbol{\alpha})$ and $q(x, \boldsymbol{\tau})$.

An m-parameter exponential family $\{f(x, \zeta)\}$ can be shown to form a Riemannian manifold[6] with the real parameter vector ζ uniquely defining a density $f(x, \zeta)$ described by the Fisher information matrix $I(\zeta)$,

$$I_{ij}(\zeta) = E_f \left(\frac{\partial \log f(x, \zeta)}{\partial \zeta_i} \frac{\partial \log f(x, \zeta)}{\partial \zeta_j} \right), \tag{5.8}$$

which may be treated as the Riemann metric tensor. Recall that the Fisher information matrix[5] provides a measure of information contained in the random variable $X \sim f(x, \zeta)$, about ζ. As such, an m-parameter exponential family defines two dual manifolds, $\mathscr{P}_m$ with its m natural parameters and $\mathscr{M}_m$ with its moment parameters.

5.2.4 Divergence distance

Assume $p(x, \boldsymbol{\alpha})$ and $p'(x, \boldsymbol{\alpha}')$ are two PDFs from an m-parameter exponential family with their respective natural parameters $\boldsymbol{\alpha}$ and $\boldsymbol{\alpha}'$, moment parameters $\boldsymbol{\tau}$ and $\boldsymbol{\tau}'$, and associated potential functions $\psi_m(\boldsymbol{\alpha})$, $\phi(\boldsymbol{\tau})$ and $\psi_m(\boldsymbol{\alpha}')$, $\phi(\boldsymbol{\tau}')$. Then, their divergence distance $D(p\|p')$ may be written in terms of their potential functions as

$$D(p\|p') = E_p \log(p/p') = E_p \log p - E_p \log p' = \phi(\boldsymbol{\tau}) + \psi_m(\boldsymbol{\alpha}') - \langle \boldsymbol{\alpha}', \boldsymbol{\tau} \rangle. \tag{5.9}$$

5.2.5 Information projection

Let $p_m(x, \boldsymbol{\alpha}) \in \mathscr{P}_m$, then its projection to the k-dimensional subspace $\mathscr{P}_k$ is defined by a PDF from $\mathscr{P}_k$ as

$$p_k(x, \boldsymbol{\beta}) = \arg\min\{D(p\|q) : q(x) \in \mathscr{P}_k\}. \tag{5.10}$$

Now, let $\mathscr{M}_k(p_m)$ denote the set of all densities from an exponential family having the same set of k moments as p_m on a set of k statistics, a subset of the set of m statistics that generates p_m. The projection $p_k(x, \boldsymbol{\beta})$ belongs to $\mathscr{M}_k(p_m)$ and maximizes the entropy functional over all PDFs from this set,[6] i.e.,

$$p_k(x, \boldsymbol{\beta}) = \arg\max\{H(q) : q(x) \in \mathscr{M}_k(p_m)\}. \tag{5.11}$$

We discuss the consequence of this result in the next section.

5.3 Properties of Entropy Estimates

Recall that the ME estimate of $f(x)$, satisfying a given set of moment constraints, is given by[1]

$$p_m(x, \alpha) = e^{\langle \boldsymbol{\alpha}, \mathbf{T}(\mathbf{x}) \rangle - \psi(\boldsymbol{\alpha})}. \tag{5.12}$$

The parameters $\{\alpha_k\}$ are chosen as functions of the given moments so that $p_m(x, \alpha)$ maximizes the entropy functional (objective function), subject to the moment constraints. The problem of this constrained optimization may be solved via the method of Lagrange multipliers. Note that this estimate is a member of an m-parameter canonical exponential family with natural parameters α_k maximizing $L(\boldsymbol{\alpha}, \boldsymbol{\tau}) = \langle \boldsymbol{\alpha}, \boldsymbol{\tau} \rangle - \psi(\boldsymbol{\alpha})$. The existence of a solution vector $\boldsymbol{\alpha}$ is guaranteed by the concavity of $L(\boldsymbol{\alpha}, \boldsymbol{\tau})$.

5.3.1 Maximum entropy and maximum likelihood

When the prescribed m moments are sample moments $\hat{\tau}_k$ based on an independent sample $X = \{x_1, x_2, \ldots, x_n\}$ from an unknown density $f(x)$, the ME estimate is obtained as a density $p(x, \boldsymbol{\alpha})$ from an m-parameter exponential family whose parameters maximize the empirical log-likelihood function. However, this is precisely how the parameters of a density $p(x, \boldsymbol{\alpha})$ from an m-parameter exponential family are estimated by the ML principle. As such, the ME estimation technique may be viewed as a *restricted* ML estimation method, where $f(x)$ is restricted to be a density $p(x, \boldsymbol{\alpha})$ from an m-parameter exponential family where m specifies the number of moment constraints.

5.3.2 Invariance property

The proof of the following theorem, which establishes the invariance of the ME estimation method under projection, follows directly from known results on information projection.[6]

Theorem 5.3.1. Let $p_m(x, \boldsymbol{\alpha})$ be a ME estimate belonging to an mth-order exponential family. Then its projection $p_k(x, \boldsymbol{\beta})$ onto its k dimensional subspace $\mathscr{P}_k$ is also a ME estimate.

ML estimation has an analogous, but stronger, invariance property: the MLE is invariant under a one-to-one mapping of its parameter space.

5.4 Estimation Error Bounds

We now establish bounds on the error associated with ME estimates. Two cases are considered. First, when the number of available moments exceeds the true order of the distribution to be estimated, the problem is overspecified. Conversely, when the available moments are fewer in number than the order of the true distribution, the problem is underspecified or *ill posed*.

5.4.1 Overspecified models

When the unknown probability model for an uncertain system is a density from an n-parameter exponential family and moments are known for $m(\geq n)$ linearly independent statistics, the estimation problem is overdetermined; the question arises as to whether the ME method correctly determines the underlying model. We show that the ME method indeed provides the exact model, a density from an n-parameter exponential family.

Assume the availability of m moments constructed from a sequence of linearly independent statistics, $\{T_1(x), \ldots, T_n(x), U_1(x), \ldots, U_{m-n}(x)\}$, from an unknown density belonging to an n-parameter exponential family with a minimal representation:

$$p(x, \boldsymbol{\alpha}) = e^{\langle \boldsymbol{\alpha}, \mathbf{T(x)} \rangle - \psi_n(\boldsymbol{\alpha})} \quad \text{where} \quad \psi_n(\boldsymbol{\alpha}) = \log \int e^{\langle \boldsymbol{\alpha}, \mathbf{T(x)} \rangle} dx < \infty. \quad (5.13)$$

We are now ready to state the following theorem:

Theorem 5.4.1. Let $\mathbf{T(x)}$ and $\mathbf{U(x)}$ be two vectors of linearly independent statistics, and let $\boldsymbol{\tau} = E(\mathbf{T(x)})$ and $\mathbf{u} = \mathbf{E(U(x))}$ be the moments generated by these statistics used by the ME method to determine an exponential representation

$$p(x, \tilde{\boldsymbol{\alpha}}) = e^{\langle \tilde{\boldsymbol{\alpha}}, \tilde{\mathbf{T}}(x) \rangle - \psi_n(\tilde{\boldsymbol{\alpha}})} \quad \text{where} \quad \tilde{\mathbf{T}} = [\mathbf{T^T}, \mathbf{U^T}]^\mathbf{T}, \quad (5.14)$$

and let $p(x, \boldsymbol{\alpha})$ be an exact minimal representation of the unknown density. Then, maximizing $L(\tilde{\boldsymbol{\alpha}}, (\boldsymbol{\tau}, \mathbf{u}))$ is equivalent to maximizing $L(\boldsymbol{\alpha}, \boldsymbol{\tau}) = \langle \boldsymbol{\alpha}, \boldsymbol{\tau} \rangle - \psi_n(\boldsymbol{\alpha}) = \phi(\boldsymbol{\tau})$, the Legendre transform of $\psi_n(\boldsymbol{\alpha})$.
Proof: See Refs. 7 or 8.

We have therefore shown that the ME estimate for an overdetermined system yields an exact estimate of the correct order; i.e., there is no error in the estimate.

5.4.2 Underspecified models

Typically, the number of parameters required for a minimal exponential representation of a density of any practical significance is finite. Assume now that for an unknown PDF from an n-parameter exponential family, we have available moments of the $m(< n)$ statistics $\mathbf{T(x)} = [\mathbf{T_1(x)} \ldots \mathbf{T_m(x)}]$. Let $\{\mathbf{T(x), U(x)}\}$ be n linearly independent statistics generating an n-parameter exponential family member, where

$$\mathbf{U(x)} \triangleq [\mathbf{U_1(x)} \ldots \mathbf{U_{n-m}(x)}].$$

Then, for some $\boldsymbol{\alpha}$ and $\boldsymbol{\beta}$, the unknown density $f(x)$ from an n-parameter exponential family has a minimal exponential representation:

$$f(x) = p_n(x, \boldsymbol{\alpha}, \boldsymbol{\beta}) = e^{\langle \boldsymbol{\alpha}, \mathbf{T(x)} \rangle + \langle \boldsymbol{\beta}, \mathbf{U(x)} \rangle - \psi(\boldsymbol{\alpha}, \boldsymbol{\beta})}. \quad (5.15)$$

Again, define $\boldsymbol{\tau} = E_f(\mathbf{T})$ and $\mathbf{u} = \mathbf{E_f(U)}$. Then, for a given moment vector $\boldsymbol{\tau}^0$, the ME estimate is obtained as

$$p_m(x, \boldsymbol{\alpha}^0) = e^{\langle \boldsymbol{\alpha}^0, \mathbf{T} \rangle - \psi(\boldsymbol{\alpha}^0)}, \quad (5.16)$$

where $\boldsymbol{\alpha}^0$ maximizes $L(\boldsymbol{\alpha}, \boldsymbol{\tau}^0)$. Given a $\boldsymbol{\tau}^0$, the partial Legendre transform of $\psi(\boldsymbol{\alpha}, \boldsymbol{\beta})$ is given as

$$\phi(\boldsymbol{\tau}^0, \boldsymbol{\beta}) = \max_{\boldsymbol{\alpha}} [\langle \boldsymbol{\alpha}, \boldsymbol{\tau}^0 \rangle - \psi(\boldsymbol{\alpha}, \boldsymbol{\beta})] = \langle \bar{\boldsymbol{\alpha}}, \boldsymbol{\tau}^0 \rangle - \psi(\bar{\boldsymbol{\alpha}}, \boldsymbol{\beta}) \quad (5.17)$$

where $\bar{\alpha}$ satisfies the gradient equation $\tau^0 = \nabla_\alpha \psi(\alpha, \beta)$. Note that $\phi(\tau^0, \beta)$ is a concave function of β. Thus, $\phi(\tau^0, \beta)$ has a unique maximum at $(\bar{\alpha}, \bar{\beta})$ and $\phi(\tau^0, \bar{\beta}) > \phi(\tau^0, \beta)$ for any other β. Now, let $u^0 = \nabla_\beta \psi(\bar{\alpha}, \bar{\beta})$, then we claim that the PDF

$$f_n(x, \bar{\alpha}, \bar{\beta}) = e^{\langle \bar{\alpha}, \mathbf{T} \rangle + \langle \bar{\beta}, \mathbf{U} \rangle - \psi(\bar{\alpha}, \bar{\beta})} \tag{5.18}$$

from an n-parameter canonical exponential family has the maximum divergence distance from the estimate $f_n(x, \alpha^0)$.

Theorem 5.4.2. Of all PDFs from an n-parameter canonical exponential family, $f_n(x, \bar{\alpha}, \bar{\beta})$ has the maximum divergence distance from the ME estimate $f_m(x, \alpha^0)$ computed from the information $\tau^0 = E_f(\mathbf{T})$.
Proof: See Refs. 7 or 8.

Since $(\bar{\alpha}, \bar{\beta})$ maximizes $\langle \alpha, \tau^0 \rangle + \langle \beta, u \rangle - \psi(\alpha, \beta)$, $D(f_m \| f_n)$ is maximized at $f_n(x, \bar{\alpha}, \bar{\beta})$. Thus, $D(f_n(x, \alpha, \beta) | f_m) \leq D(f_n(x, \bar{\alpha}, \bar{\beta}) \| f_m)$, for all β and $\alpha = \alpha(\tau^0, \beta)$. Therefore, we have proved that there exists a density from an exponential family of order n that has the same m prescribed moments and largest divergence distance from the ME estimate, thereby bounding the error of any ME estimate.

5.5 Data-Driven Estimation

Assume a large sample $\mathbf{X} = \{\mathbf{x_1}, \mathbf{x_2}, \ldots, \mathbf{x_N}\}$ of N independent observations is available from a density of an m-parameter exponential family with minimal representation

$$p(x, \alpha) = e^{\langle \alpha, \mathbf{T(x)} \rangle - \psi_m(\alpha)}. \tag{5.19}$$

We may write the log-likelihood $\ell_N(\alpha | \mathbf{X})$ as

$$\ell_N(\alpha | \mathbf{X}) = \sum_i [\langle \alpha, \mathbf{T(x_i)} \rangle - \psi_m(\alpha)] = \mathbf{N}[\langle \alpha, \hat{\tau} \rangle - \psi_m(\alpha)], \tag{5.20}$$

where $\hat{\tau} = \frac{1}{N} \sum \mathbf{T(x_i)}$ is the ML estimate of $\tau = E_p \mathbf{T(x)}$.

There are several methods, including the Akaike information criterion,[9] that may be used to estimate the order of a PDF from an exponential family underlying a sample data set. However, as we have seen, the ME estimation method using the sample moments $\hat{\tau}$ as prescribed moments can provide an estimate of the order of the density that is generating the data X. Wilks's theorem[5] can be used to test the validity of this estimate.

Beginning with m sample moments $\hat{\tau}$ formed via m linearly independent statistics $\mathbf{T(x)}$, the ME estimation method finds a parameter vector $\hat{\alpha}$ that maximizes $\langle \alpha, \hat{\tau} \rangle - \psi_m(\alpha)$. As $\hat{\alpha}$ maximizes the log-likelihood $\ell_n(\alpha | \mathbf{X})$, it is also the MLE. Let α^0 be the n nonzero components of $\hat{\alpha}$ and $\mathbf{T^0(x)}$ the corresponding statistics vector. Then the ME estimate of the underlying density is given by

$$p(\mathbf{x}, \alpha^0) = e^{\langle \alpha^0, \mathbf{T^0(x)} \rangle - \psi_n(\alpha^0)}. \tag{5.21}$$

So, we have

$$ND(p(x, \hat{\boldsymbol{\alpha}}) \| p(x, \boldsymbol{\alpha}^0)) = [\ell_N(\hat{\boldsymbol{\alpha}} | \mathbf{X}) - \ell_{\mathbf{N}}(\boldsymbol{\alpha}^0 | \mathbf{X})]. \qquad (5.22)$$

If the density generating $\mathbf{X}$ is indeed from an m-parameter exponential family, then by Wilks's theorem, we have approximately

$$ND(p(x, \hat{\boldsymbol{\alpha}}) \| p(x, \boldsymbol{\alpha}^0)) \sim \chi^2_{m-n}. \qquad (5.23)$$

This last result may be used either to construct a test of significance or to provide a confidence interval for the density estimate in terms of divergence.

5.6 Conclusion

We have derived an upper bound for error in approximating an unknown density function $f(x)$ by its ME estimate based on m moment constraints, obtained as a PDF $p(x, \boldsymbol{\alpha})$ from an m-parameter exponential family. The error bound will help us decide if the number of moment constraints is adequate for modeling the uncertainty in the system under study. In turn, this allows one to establish confidence intervals on an estimate of some function of the random variable $\mathbf{X}$ given the known moments. We have also shown how, when working with a large sample of independent observations, instead of precisely known moment constraints, we can define a density from an exponential family to model the uncertainty of the underlying system with measurable accuracy. In this case, we have also established a relationship to the MLE.

References

1. T. Cover and J. Thomas, *Elements of Information Theory*, John Wiley & Sons, New York, NY (2006).

2. V. Arnold, *Mathematical Methods of Classical Mechanics*, Springer, New York, NY (1989).

3. R. Courant and D. Hilbert, "Methods of mathematical physics, Vol. I," *Physics Today* **7**, 17 (1954).

4. O. Barndorff-Nielsen, *Exponential Families Exact Theory*, Aarhus Universitet, Matematisk Institut, Aarhus, Denmark (1970).

5. P. Bickel and K. Doksum, *Mathematical Statistics*, Prentice-Hall, Upper Saddle River, NJ (2006).

6. S. Amari, "Information geometry on hierarchy of probability distributions," *IEEE Transactions on Information Theory* **47**(5), 1701–1711 (2001).

7. P. Basu, *Estimation and System Modeling Using Information Theoretic Measures*. PhD thesis, Tufts University (2007).

8. J. Noonan and P. Basu, "On estimation error using maximum entropy density estimates," *Kybernetes* **36**(1), 52–64 (2007).

9. H. Akaike, "A new look at the statistical model identification," *IEEE Transactions on Automatic Control* **19**(6), 716–723 (1974).

Part III

Point Estimation: Signal Restoration

Chapter 6
The Stabilizing-Functional Approach to Regularization

6.1 Introduction

Signal formation is often represented by the standard linear model:

$$\mathbf{y} = \mathbf{Hx} + \mathbf{n}, \tag{6.1}$$

where $\mathbf{x}$ and $\mathbf{y}$ are the original and observed signals, respectively, and $\mathbf{n}$ is the additive noise due to the measuring device. Signal restoration is the process of inferring the best estimate for the target signal $\mathbf{x}$ given the observed signal $\mathbf{y}$ and some prior knowledge, if available, about the target.

An inverse problem is said to be ill posed when direct inversion does not ensure the existence, uniqueness, and stability of a solution. Signal restoration generally belongs to this class of problems, and regularization theory formulates how solutions may be found for such ill-posed problems.[1-3] One method for developing such solutions is the stabilizing-functional approach wherein the ill-posed problem is recast as a constrained minimization of a chosen functional, which is called a stabilizing functional.

The first regularization techniques for signal restoration were often based on mean-square norms. It has been shown that such constrained least-squares approaches are related to the stabilizing-functional approach via quadratic functionals of a special form.[2] Here, we will be concerned with using the nonquadratic functionals typically encountered in information theory.

6.2 Regularization Theory

To begin our discussion, we restate the definitions of well-posed and ill-posed inverse problems.

Definition 6.2.1. Determining $\mathbf{x} \in \mathbf{X}$ from the observations $\mathbf{y} \in \mathbf{Y}$ is said to be a well-posed inverse problem on the pair of metric spaces $\{(\mathbf{X}, \rho_{\mathbf{x}}), (\mathbf{Y}, \rho_{\mathbf{y}})\}$ in the *Hadamard sense* if the following three conditions are satisfied:

1. Existence – For each $\mathbf{y} \in \mathbf{Y}$ there exists a solution $\mathbf{x} \in \mathbf{X}$.

2. Uniqueness – The solution is unique, i.e., $\mathbf{Hx_1} = \mathbf{Hx_2}$ if and only if $\mathbf{x_1} = \mathbf{x_2}$.

3. Stability – The problem is stable on the spaces $\{(\mathbf{X}, \rho_\mathbf{x}), (\mathbf{Y}, \rho_\mathbf{y})\}$, which is implied by the existence of a continuous inverse $\mathbf{H}^{-1}$ on $\mathbf{Y}$.

Definition 6.2.2. A problem is defined as being an ill-posed problem when direct inversion does not yield a guaranteed unique and stable solution and when bounded variations in the input cause unbounded variations in the solution.

Solutions obtained via direct inversion of the observation apparatus [Eq. (6.1)] are typically unstable; there is no gain in trying to perform better than noise and uncertainty allow. Regularization theory recognizes this and recasts an ill-posed problem as a related well-posed problem whose solution is stable, unique, and in some sense a close approximation of the solution to the original ill-posed problem.[2] Key to this approach is the selection of a suitable *regularizing operator*, defined formally as:

Definition 6.2.3. An operator $R(\mathbf{y}, \alpha)$ depending on the *regularization parameter* α is called a *regularizing operator* for Eq. (6.1) in a neighborhood of $\mathbf{y_T}$, the noiseless observations, if

1. there exists a $\delta_1 > 0$ such that the operator $R(\mathbf{y}, \alpha)$ is defined for every $\alpha > 0$ and every $\mathbf{y} \in \mathbf{Y}$ for which $\rho_y(\mathbf{y}, \mathbf{y_T}) \leq \delta \leq \delta_1$,

2. there exists a function $\alpha = \alpha(\delta)$ such that for every $\varepsilon > 0$, there is a $\delta(\varepsilon) \leq \delta_1$ such that $\mathbf{y} \in \mathbf{Y}$ and

$$\rho_y(\mathbf{y_T}, \mathbf{y}_\delta) \leq \delta(\varepsilon) \Rightarrow \rho_\mathbf{x}(\mathbf{x_T}, \mathbf{x}_\alpha) \leq \varepsilon,$$

   where $\mathbf{x}_\alpha = \mathbf{R}(\mathbf{y}_\delta, \alpha(\delta))$ is called the *regularized solution*, and the regularizing operator is not necessarily unique.

Every regularizing operator defines a stable method of approximating the solution of Eq. (6.1) provided the choice for α is consistent with the variability δ of the initial data. Under this approach, finding a stable approximate solution to Eq. (6.1) entails finding regularizing operators and determining the regularizing parameter from supplementary information pertaining to the problem.

6.3 The Stabilizing-Functional Approach

Of the many methods used to generate valid regularizing operators, most popular is the *stabilizing-functional approach*, presented here following the development of Karayiannis[2] and Katsaggelos.[4]

The stabilizing-functional approach arrives at a solution to the ill-posed problem by formulating it as a constrained minimization of a specific functional, called the stabilizing functional. While the mathematics of minimization is quite well established, the choice of a particular functional requires some insight and a willingness to do heuristic experimentation.

The simplest technique in this class is the one obtained by minimizing the norm itself. This is equivalent to the CLSR described earlier. In fact, Karayiannis and Venetsenapoulos[2] have shown that any of the CLSR algorithms may be obtained by using a properly chosen quadratic stabilizing functional.

Another important aspect of this approach is that the ME solution to the image restoration problem can also be formulated within this paradigm by using the negative entropy functional as the stabilizing functional, i.e., maximize:

$$-\mathbf{x}^T \log_e (\mathbf{x}) \tag{6.2}$$

subject to:

$$\| \mathbf{y} - \mathbf{H} \cdot \mathbf{x} \| = \| \mathbf{n} \| . \tag{6.3}$$

Some straightforward manipulation[5] yields the following solution:

$$\mathbf{x} = exp\left(-\mathbf{1}_c - 2\lambda \mathbf{H}^T (\mathbf{y} - \mathbf{H} \cdot \mathbf{x})\right), \tag{6.4}$$

where λ is a Lagrange multiplier, and $\mathbf{1}_c$ is a column vector all of whose elements equal 1. Trussell[5] has shown that one form of the ME solution is a special case of the MAP solution. It follows that for some cases at least the stabilizing-functional approach leads to a MAP estimate. We take care to note that regularization with information measures in a non-probabilistic setting is not without precedent. In Lee[6] and Donoho,[7] the KL information divergence measure is extended to operate on nonnegative matrices. Matrix factorization methods relying on such an approach have been successfully applied to problems such as image classification,[8] document clustering,[9] and music transcription.[10]

For the minimization process to be stable it is necessary that only functionals possessing a unique minimum be chosen as stabilizing functional. In defining the following criteria that a functional must satisfy to be chosen as a stabilizing functional we follow the definition in Ref. 2.

Let $\Omega(x)$ be a continuous nonnegative functional defined on a subspace $\hat{S}$ of a space $\mathcal{R}$. $\Omega(\mathbf{x}) \leq \epsilon$ is a stabilizing functional if

- The solution $\mathbf{x}^*$ to $\mathbf{y} = \mathbf{H} \cdot \mathbf{x} + \mathbf{n}$ belongs to the domain of $\Omega(x)$.

- For every positive ϵ, the set of elements $\mathbf{x}$ of $\hat{S}$ for which $\Omega(\mathbf{x}) \leq \epsilon$ is a compact subset in $\mathcal{S}$.

With these definitions, the stabilizing-functional approach is formulated as

Definition 6.3.1. The regularization theory stabilizing functional using the least-squares norm (RTSFLS)* projection is that $x^* \in \mathcal{S}$, which minimizes the functional

---

*See Appendix A for a discussion of projections.

$\Omega\left(f\right)$ and is consistent with the following constraint:

$$\parallel \mathbf{y} - \mathbf{H} \cdot \mathbf{x}^* \parallel = \parallel n \parallel, \tag{6.5}$$

where the norm $\parallel \cdot \parallel$ is the L_2 norm.

While the above statement formulates the approach, it tells us nothing about the choice of the exact functional itself, except for imposing some general constraints on the choice. Various researchers have used different stabilizing functionals for the solution of the inverse problem. Among the more popular functionals are the class of quadratic functionals introduced by Karayiannis and Venetsanopoulos. Other functionals such as the Itakura–Saito distance measure for spectral estimation and Csiszar's I-divergence function have also been used. See Appendix A for a brief but broad coverage of the various distance measures that are commonly used. There is thus a wide variety of choice in the selection of the stabilizing functional.

Some researchers have suggested that conformance with the principle of directed orthogonality[†] is a desirable feature in a stabilizing functional. While in a strictly formal sense, it is necessary that a measure obeys this principle to be consistent with the idea of *best approximations*, it is of importance to note here that the only essential requirement is the existence of a minimum. Other conditions may be imposed so as to restrict the choice of a functional, but such conditions are *not essential* for the existence of a stable, unique solution. The existence of a minimum is guaranteed if the functional is convex.

For a detailed consideration of various quadratic stabilizing functionals, see Ref. 2. Katsaggellos[4] describes an interesting variant of this approach based on set theoretic considerations.

6.3.1 Mathematical formulation

By choice, a stabilizing functional Ω possesses a minimum. Using the techniques of variational calculus, this constrained minimization is rewritten as an unconstrained minimization of a smoothing functional $M\left(\mathbf{x}, \mathbf{x}_p\right)$. The problem then is that of determining the $\mathbf{x}^* \in \mathcal{S}_1$ that yields the minimum value of the functional. Thus, we would like to minimize

$$M\left(\mathbf{x}, \mathbf{x}_p\right) = \Omega\left(\mathbf{x}, \mathbf{x}_p\right) + \lambda \parallel \mathbf{y} - \mathbf{H} \cdot \mathbf{x} \parallel . \tag{6.6}$$

Often it is necessary to use iterative techniques to obtain a solution to the minimization problem.

6.4 Conclusion

Solutions to ill-posed problems obtained via direct inversion are often unstable and hence unreliable. Regularization theory offers a workaround by sacrificing a small

---

[†]See Appendix C.

amount of precision for stability. Typically, ill-posed problems are regularized using quadratic functionals, but the *stabilizing-functional approach* to regularization guarantees that using any convex functional is also effective. In particular, certain functionals arising in information theory, such as entropy, may be used to generate solutions of interest in a particular setting.

References

1. N. Zuhair and M. Nashed, "Lecture notes in mathematics," in *Functional Analysis Methods in Numerical Analysis*, Proceedings of a Special Session held at the Annual Meeting of the American Mathematical Society (1977).

2. N. Karayiannis and A. Venetsanopoulos, "Regularization theory in image restoration – the stabilizing functional approach," *IEEE Transactions on Acoustics, Speech and Signal Processing* **38**(7) (1990).

3. A. Tikhonov and V. Arsenin, *Solutions of Ill-Posed Problems*, V.H. Winston & Sons, Washington, DC (1977).

4. A. Katsaggelos, J. Biemond, R. Schafer, and R. Mersereau, "A regularized iterative image restoration algorithm," *IEEE Transactions on Signal Processing* **39**(4), 914–929 (1991).

5. H. Trussell, "The relationship between image restoration by the maximum a posteriori method and a maximum entropy method," *IEEE Transactions on Acoustics, Speech and Signal Processing* **28**(1), 114–117 (1980).

6. D. Lee and H. Seung, "Algorithms for non-negative matrix factorization," *Advances in neural information processing systems* **13** (2001).

7. D. Donoho and V. Stodden, *When does non-negative matrix factorization give a correct decomposition into parts?* MIT Press, Cambridge, MA (2003).

8. D. Guillamet, J. Vitrika, and B. Schiele, "Introducing a weighted non-negative matrix factorization for image classification," *Pattern Recognition Letters* **24**(14), 2447–2454 (2003).

9. W. Xu, X. Liu, and Y. Gong, "Document clustering based on non-negative matrix factorization," in *Proceedings of the 26th Annual International ACM SIGIR Conference on Research and Development in Informaion Retrieval*, 267–273, ACM (2003).

10. P. Smaragdis and J. Brown, "Non-negative matrix factorization for polyphonic music transcription," in *Applications of Signal Processing to Audio and Acoustics, 2003 IEEE Workshop on.*, 177–180, IEEE (2004).

Chapter 7

Applying Information-Theoretic Distances and the GMF

7.1 Introduction

In Chapter 6, the stabilizing-functional approach to regularizing the solution to the inverse problem was described. In the following we lay down the framework for presenting the generalized mapping function (GMF). The signal space is symbolized by $\mathcal{S}$. In the absence of any constraint on the solution, the estimated signal is the vector $\mathbf{x}^* \in \mathcal{S}$ that minimizes the stabilizing functional. The addition of a constraint(s) effectively forces the solution to belong to a subspace $\mathcal{S}_1 \in \mathcal{S}$. The search for the true solution is now restricted to the subspace $\mathcal{S}_1$; the more constraints imposed, the smaller the space over which the search for the true solution occurs. Imposing additional constraints will speed up the algorithm, but unless the constraints are well grounded in physics, their use could lead to spurious solutions. We will use the noise norm constraint, which uses the noise variance. That is, SNR information is the only information about the noise assumed to be available.

Since we are concerned with finding solutions to problems involving measurements, it is reasonable to assume that the signals are time limited and that the spectra are bandlimited. By choice we will restrict our discussion to the space of nonnegative functions.

7.1.1 Definition of the signal space

Let $\mathbf{x}$ be a sampled signal/spectra or a sampled image represented in lexicographic order. By the above discussion the elements of $\mathbf{x}$ have values in the set of nonnegative real numbers. The set of nonnegative real numbers is symbolized by $\mathbb{R}^+$. Further, for the purposes of deriving the GMF, we will identify the normalized signals as their PDFs. This approach has been earlier used by Frieden[1] in deriving a ME-based solution. The normalized signals are also represented by $\mathbf{x}$ since no confusion seems likely by this dual usage. Thus the signals are mapped into the space $\mathcal{S}$ which is defined as

$$\mathcal{S} = \left\{ \mathbf{x} \mid \sum_{i=1}^{i=k} \mathbf{x}[i] = 1 \right\}. \tag{7.1}$$

The mapping need not always be an invertible one. This is especially true if an iterative technique is used to locate the solution since some information about the exact energy or intensity at each sample point may be lost during such a process.

The norm (distance measure) on $\mathcal{S}$ is defined by $\| \, . \, \|$ such that

$$\| \, \mathbf{x} \, \| = \sqrt{\frac{1}{k} \sum_{i=1}^{k} \mathbf{x}^2 \, [i]}. \tag{7.2}$$

With this notation, we make the following claim.

Claim 7.1.1. The set $\mathcal{S}$ is a subset of $\mathcal{S}_c$

$$\mathcal{S} \subset \mathcal{S}_c = \left\{ \mathbf{x} \, \| \| \, \mathbf{x} \, \| \leq \sqrt{\frac{1}{k}} \right\}. \tag{7.3}$$

Proof:
The proof is straightforward. Let $\mathbf{x} \in \mathcal{S}$, where the set $\mathcal{S}$ is given by Eq. (7.1). Let $\mathbf{x} \, [i]$ denote the ith element of the vector $\mathbf{x}$. Then

$$(\mathbf{x} \, [1] + \ldots + \mathbf{x} \, (k))^2 = \mathbf{x}^2 \, [1] + \ldots + \mathbf{x}^2 \, [k] + 2 \sum_{i=1}^{i=k} \sum_{j=1, j \neq i}^{j=k} \mathbf{x} \, [i] \, \mathbf{x} \, [j] \tag{7.4}$$

Due to the fact that all of the elements of the vector $\mathbf{x}$ are nonnegative, it follows from the above equation that

$$\mathbf{x}^2 \, [1] + \ldots + \mathbf{x}^2 \, [k] \leq (\mathbf{x} \, [1] + \ldots + \mathbf{x} \, (k))^2 = 1, \tag{7.5}$$

which may be rewritten as

$$\sum_{i=1}^{i=k} \mathbf{x}^2 \, [i] \leq 1.$$

Therefore

$$\frac{1}{k} \sum_{i=1}^{i=k} \mathbf{x}^2 \, [i] \leq \frac{1}{k}, \tag{7.6}$$

and finally

$$\sqrt{\frac{1}{k} \sum_{i=1}^{i=k} \mathbf{x}^2 \, [i]} \leq \sqrt{\frac{1}{k}}. \tag{7.7}$$

This proves that if $\mathbf{x} \in \mathcal{S}$ then $\mathbf{x} \in \mathcal{S}_c$ and thus $\mathcal{S} \subset \mathcal{S}_c$.

From Eq. (7.3), it follows very simply that $\mathcal{S}_c$ is a closed set.* We define the distance between two functions (vectors) in $\mathcal{S}_c$ as

$$d \, (\mathbf{x}, \mathbf{y}) = \| \, \mathbf{x} - \mathbf{y} \, \|. \tag{7.8}$$

---

*See Appendix C for a definition of closedness.

This distance is the metric on $\mathcal{S}_c$, and with this definition we claim that $(\mathcal{S}_c, d)$ is a closed metric space. We now state without proof the following basic theorem from analysis.[2]

Theorem 7.1.1. Every Cauchy sequence in $\mathbb{R}^k$ converges[†] to a point in $\mathbb{R}^k$.

Additionally, the following theorem holds true:

Theorem 7.1.2. Let $(\mathcal{M}, d)$ be a complete metric space. Suppose $\mathcal{N} \subset \mathcal{M}$ is a closed subset, then $\mathcal{N}$ is a complete metric space with the metric on $\mathcal{N}$ being the same as the metric on $\mathcal{M}$.

Noting that $\mathcal{S}_c \subset \mathbb{R}^k$, by the closedness of $\mathcal{S}_c$ and by the previous two theorems we make the following claim:

Claim 7.1.2. $(\mathcal{S}_c, d)$ is a complete metric space.

Once again, we note that $\mathcal{S}$ is a closed subset of $\mathcal{S}_c$. Therefore, by Theorems **7.1.1** and **7.1.2** we make the following claim.

Claim 7.1.3. $(\mathcal{S}, d)$ is a complete metric space.

The completeness of $\mathcal{S}$ guarantees that if there is a convergent sequence in $\mathcal{S}$, then the limit point of that sequence is contained in $\mathcal{S}$. We now state a fundamental theorem from analysis that will be used in discussing the convergence of the GMF. The theorem, known as the contraction mapping theorem, specifies the conditions under which an iteration converges. This is central to the analysis of iterative methods since the first requirement for an iteration system to be of any use is that it always converge. The other important requirement is that this convergence should be to a unique solution, i.e., the limit point of the iteration should be independent of initial conditions. The power of the contraction mapping theorem comes from the fact that it guarantees the uniqueness of the solution if the specified conditions are met.

Theorem 7.1.3 (Contraction mapping theorem). Given the complete metric space $\mathcal{M}$, let $\phi : \mathcal{M} \to \mathcal{M}$ be a given mapping. Assume that there is a constant $\alpha, 0 \leq \alpha$ such that

$$d(\phi(x), \phi(y)) \leq \alpha \, d(x, y) \quad \forall x, y \in \mathcal{M}, \tag{7.9}$$

then there is a unique fixed point for ϕ, i.e., a point x_* such that $\phi(x_*) = x_*$. Also, if $x_0 \in \mathcal{M}$ and if we define

$$x_1 \triangleq \phi(x_0), \; x_2 \triangleq \phi(x_1), \; \ldots, \; x_{n+1} \triangleq \phi(x_n),$$

then

$$\lim_{n \to \infty} x_n = x_*.$$

[†]Convergence is defined in Appendix C.

In this section, the structure for presenting and analyzing the performance of the GMF has been built. In the next section we discuss the set of feasible solutions.

7.1.2 The set of feasible solutions

The estimate $\hat{x}$ of the true signal x is to be chosen from all k-tuples $x \in \mathbb{R}_+^k$, such that it satisfies some prespecified conditions. The constraint that we choose to impose is that the proposed solution $\hat{x}$ be such that it satisfies

$$\| \mathbf{y} - \mathbf{H} \cdot \hat{x} \| = \| \mathbf{n} \| . \tag{7.10}$$

This restricts the set of possible solutions to functions that are consistent with the above constraint.[‡] This set is referred to as the set of feasible solutions. As a practical matter, this implies that the iteration should be stopped at iteration p such that

$$\| \mathbf{y} - \mathbf{H} \cdot \mathbf{x_p} \| = \| \mathbf{n} \| . \tag{7.11}$$

7.2 Mutual Information Signal Restoration

The use of information-theoretic criteria for signal restoration was pioneered by, among others, Burg[4] and Frieden.[1] Both Burg and Frieden used the extremum of the entropy functional as a method to ensure uniqueness of the solution. This approach requires the PDF of the observed process. However, such information is usually not available; the mean and variance of the noise are the only quantities known or estimated. Burg and Frieden chose different approaches to get around this problem. Frieden normalized the signal so that it summed to unity. This is possible under the assumption of nonnegativity. This normalized signal was identified as the PDF of the signal or image. While no theoretical justification has been proffered for this approach, experimental results have endowed an empirical validity to the scheme.

Noonan and Tzannes first considered entropy and MI as a principle for regularizing inverse problems in Refs. 5 and 6. Noonan and Marcus[7,8] applied the mutual information measure (MIM) to the modeling of a communications channel, while Noonan and Achour[9,10] investigated the application of this functional to signal restoration. Achour[11] investigated in detail various other entropic distances such as the KL measure and the Itakura–Saito distance and studied the use of the MIM as a stabilizing functional for use in signal restoration. In this work, we derive this restoration scheme and interpret it as a GMF, which encompasses a broad class of iterative signal-restoration algorithms. The convergence of the mapping is analyzed, and convergence conditions are obtained for the general formulation. The conditions are applied to a particular form of the mapping. In the next chapter, we

---

[‡]Note that the constraint actually defines an ellipsoidal set under the mean-square norm. Katsaggellos[3] has proposed using the center of this ellipsoid as the solution.

present three well-known techniques that are derived from other considerations and show that these are specific forms of the GMF.

7.2.1 Mutual information

A measure of the interdependence of two processes is MI. In the case of a channel, it measures the relation between the input and output of the system. Symbolizing MI by Ω_{MI} and the PDF of a process x by $p_X(x)$, we have the following equality:

$$\Omega_{MI}(x, y) = \int_x \int_y p_{X,Y}(x, y) \ln \left[\frac{p_{X,Y}(x, y)}{p_x(x) p_Y(y)} \right] dx dy. \tag{7.12}$$

For any two specified processes, the MI functional evaluates to a specific number. This number may be interpreted as a measure of the difference between the two processes. From the above formulation, it is easy to verify that the MI functional evaluates to 0 if the two processes are independent.

7.2.2 Minimizing the stabilizing functional

The GMF is derived using the MIM as the stabilizing functional. This is achieved by the constrained minimization of the MI functional with a mean-squared-error constraint based on the noise variance. In the following, we briefly depart from the matrix notation and adopt the classical function representation. This is not done due to a sense of boredom with matrices but because it is felt that this change in notation will bring about notational convenience.[§] Mathematically the problem is stated as:
Minimize:

$$\Omega(x, y) = \sum_y \sum_x p_{Y,X}(y, x) \ln \left[\frac{p_{Y|X}(y|x)}{p_y(y)} \right] \tag{7.13}$$

subject to:

$$\frac{1}{K} \sum_{i=1}^K (y - h * x)^2 = \sigma^2. \tag{7.14}$$

This is equivalent to minimizing the following unconstrained functional:[7,11]

$$M_\alpha = \frac{1}{K} \sum_{i=1}^K (y - h * x)^2 + \alpha \Omega(x, y). \tag{7.15}$$

The above minimization yields the following form for the GMF:[11,12]

$$\mathcal{P}_x(i) = \mathcal{P}_x(i) \exp \left(\lambda x' ([u - x * h] * h_f) \right). \tag{7.16}$$

[§]Recall that the matrix notation was adopted for its compact and convenient representation.

x' is the partial of x with respect to its PDF $\mathcal{P}$. So far we have assumed that the PDF $\mathcal{P}$ is obtained by normalizing the data. This is done since specific information is generally not available, but often the actual density is given by some other function. Suppose that

$$\mathcal{P}_z = \varphi(x).$$

Different functional forms of φ clearly yield different iterative schemes. Each is the optimum MIM solution when the underlying density function for that data is actually that given by φ. It follows that we will obtain different optimum algorithms for various choices of φ.

Subsequent to the above discussion, we may rewrite Eq. (7.16) as

$$\varphi(x_{p+1}) = \varphi(x_p)exp\left(\lambda x_p'\left([u - x_p * h] * h_f\right)\right), \tag{7.17}$$

where

$$x_p' = \frac{\partial x_p}{\partial \varphi},$$

and h_f is the flipped version of the distorting function. In the matrix case, h_f will be replaced by $\mathbf{H}^T$, which represents the transpose of h. For the sake of completeness and later reference, we rewrite the above equation in matrix notation as

$$\varphi(\mathbf{x}_{p+1}) = \varphi(\mathbf{x}_p)exp\left(\lambda\mathbf{x}_p'\left(\mathbf{H}^T \cdot [\mathbf{y} - \mathbf{H} \cdot \mathbf{x}_p]\right)\right) \tag{7.18}$$

and

$$\mathbf{x}_p' = \frac{\partial \mathbf{x}_p}{\partial \varphi}.$$

Recognizing that any change in the structure of the functions above, for example, different choices for h and h_f, could potentially lead to suboptimal performance, it is worth emphasizing at this point that optimality in the context of regularization theory only implies uniqueness and stability. Thus while the aforementioned changes *could* lead to unstable behavior it is only reasonable to expect that for some particular choices of h and h_f in the above equation stable performance may be obtained. In general h_f is interpreted as a smoothing function, and a proper choice may be made heuristically. After each iteration, x_p is constrained to be positive and is bandlimited. This process can be represented as a projection operation on x, in the sense of Ref. 3, i.e.,

$$x_p = \mathbf{C}x_p. \tag{7.19}$$

Below, we refer to Eq. (7.17) as the GMF. In Chapter 8, some popular algorithms are shown to be special cases of this mapping. In the next section we will use the matrix notation to derive the condition for the convergence of the GMF.

7.2.3 Convergence of the GMF

The properties of the GMF and of the signal space $\mathcal{S}$ are presented in this section. As proved earlier, $\mathcal{S}$ is a complete metric space with the convexity property. The conditions for the convergence of the mapping in Eq. (7.17) are investigated. A strong condition is derived for the convergence of the mapping in the general form. This is followed by an example of a mapping-specific weaker condition once a particular choice, for the mapping, is made. It is established that a judicious choice of the smoothing parameter λ guarantees convergence in this particular case.

7.2.3.1 Convergence criteria—strong condition

We now proceed to derive a strong condition for the convergence of the generalized mapping. The stiffness of the condition is a natural consequence of the fact that the generality of the mapping is not diluted in the process of establishing this convergence criteria.

Rewriting the mapping in Eq. (7.17) in standard form,

$$\mathbf{x}_{p+1} = \varphi^{-1}\left(\varphi(\mathbf{x}_p)exp\left(\lambda\mathbf{x}'_p\left(\mathbf{H}^T \cdot [\mathbf{y} - \mathbf{H} \cdot \mathbf{x}_p]\right)\right)\right). \tag{7.20}$$

For convenience Eq. (7.20) is rewritten as

$$\mathbf{x}_{p+1} = \mathcal{T}(\mathbf{x}_p), \tag{7.21}$$

so that $\mathcal{T} : \mathcal{S} \to \mathcal{S}$.

Recalling that the contraction mapping theorem guarantees the convergence of the above mapping if it is shown to be a contractive mapping, we prove the following claim.

Lemma 7.2.1. $\mathcal{T} : \mathbb{R}^k \to \mathbb{R}^k$ is a contraction if

$$sup\left\{\|\mathbf{J}(\mathbf{z})\|: \mathbf{z} \in \mathcal{L}(\mathbf{x}, \mathbf{y})\right\} < 1. \tag{7.22}$$

$\mathbf{J}$ is the *Jacobian* of the map $\mathcal{T}$, and $\mathcal{L}$ is described in Definition **7.2.1** below. The *Jacobian* matrix $\mathcal{J}$ of the map $\mathcal{T}$ at a point $\mathbf{z} \in \mathbb{R}^k$ is made up of elements given by $\left\{\frac{\partial \mathcal{T}_i(\mathbf{z})}{\partial \mathbf{z}[j]}\right\}$, where $1 \leq i, j \leq k$. If the condition in Eq. (7.26) below is met, then the contraction mapping principle guarantees the existence of a unique fixed point, $\mathbf{x}^* \in \mathcal{S}$ for the iteration in Eq. (7.20) such that

$$\lim_{p \to \infty} \mathbf{x}_p = \mathbf{x}^*. \tag{7.23}$$

The generalized mean value theorem offers the structure for providing a very simple and elegant derivation of the above condition. First, we establish the convexity of $\mathcal{S}$, which is necessary for the theorem to hold.

Definition 7.2.1. Let $\mathcal{S}$ be a set such that for every $x, y \in \mathcal{S}$ the line segment $\mathcal{L}(x, y)$ joining x and y lies in $\mathcal{S}$. Then $\mathcal{S}$ is said to be convex. If $\lambda, \mu \geq 0$ are such that $\lambda + \mu = 1$ then $\lambda x + \mu y \in \mathcal{S}$.

Claim 7.2.1. $\mathcal{S}$ is convex.

Proof:

We rewrite Eq. (7.1) for convenience

$$\mathcal{S} = \left\{ \mathbf{x} \mid \sum_{i=1}^{i=k} \mathbf{x}[i] = 1 \right\}. \tag{7.24}$$

Let $\mathbf{x}, \mathbf{y} \in \mathcal{S}$ and let $\mathbf{z}$ be a point such that, for some $\beta, 0 \leq \beta \leq 1$,

$$\mathbf{z} = \beta \mathbf{x} + (1 - \beta) \mathbf{y}.$$

Then, it follows that

$$\sum_{i=1}^{i=k} \mathbf{z}[i] = \sum_{i=1}^{i=k} (\beta \mathbf{x}[i] + (1 - \beta) \mathbf{y}[i]), \tag{7.25}$$

which may be rewritten as

$$\sum_{i=1}^{i=k} \mathbf{z}[i] = \beta \sum_{i=1}^{i=k} \mathbf{x}[i] + (1 - \beta) \sum_{i=1}^{i=k} \mathbf{y}[i].$$

Since $\sum_{i=1}^{i=k} \mathbf{x}[i] = 1$, and $\sum_{i=1}^{i=k} \mathbf{y}[i] = 1$, it follows, from the definition of $\mathcal{S}$, that

$$\sum_{i=1}^{i=k} \mathbf{z}[i] = \beta \cdot 1 + (1 - \beta) \cdot 1,$$

and therefore

$$\sum_{i=1}^{i=k} \mathbf{z}[i] = \beta + 1 - \beta.$$

Thus,

$$\sum_{i=1}^{i=k} \mathbf{z}[i] = 1.$$

The above equality shows that $\mathbf{z} \in \mathcal{S}$. This concludes the proof.

In the statement of the generalized mean value theorem below, $\mathcal{T} \in C^1 [\mathbb{R}^k]$ means that the partial derivatives of $\mathcal{T}$ exist and are continuous on $\mathbb{R}^k$.

Theorem 7.2.1 (Generalized mean value theorem¶). Let $\mathcal{T}$ be a map from $\mathbb{R}^k$ to itself. That is,

$$\mathcal{T} : \mathbb{R}^k \to \mathbb{R}^k,$$

¶See Goldstein.[13]

such that the components of $\mathcal{T} \in C^1 \left[\mathbb{R}^k \right]$. Let $\mathcal{S}$ be a convex subset of $\mathbb{R}^k$. Assume that $\mathcal{T}$ possesses a *Jacobian* denoted by $\mathbf{J}$, at each point in $\mathcal{S}$. Then

$$\| \mathcal{T}\mathbf{x} - \mathcal{T}\mathbf{y} \| \leq sup \left\{ \| \mathbf{J}(\mathbf{z}) \| : \mathbf{z} \in \mathcal{L}(\mathbf{x}, \mathbf{y}) \right\} \| \mathbf{x} - \mathbf{y} \|.$$

In our case, we are interested in the mapping $\mathcal{T}$ given by Eq. (7.21). For $\mathcal{T}$ to be a contraction by Theorem **7.2.1** and the contraction mapping theorem, it is necessary to bound the supremum of the *Jacobian* so that the conditions imposed by the latter theorem are satisfied, i.e.,

$$0 < sup \left\{ \| \mathbf{J}(\mathbf{z}) \| : \mathbf{z} \in \mathcal{L}(\mathbf{x}, \mathbf{y}) \right\} < 1. \tag{7.26}$$

This requirement can be written in an explicit manner by using the definition of the norm as

$$\underset{\mathbf{z}, i, j}{max} \left| \frac{\partial \mathcal{T}_i}{\partial \mathbf{z}[j]} \right|^2 < 1. \tag{7.27}$$

That is,

$$\underset{\mathbf{z}, i, j}{max} \left| \frac{\partial \mathcal{T}_i}{\partial \mathbf{z}[j]} \right| < 1. \tag{7.28}$$

In the following section we present an example of applying this condition to a specific mapping, i.e., when $\mathcal{T}$ takes on a specific functional form.

7.2.3.2 Problem-specific weaker conditions—illustration

The nature of the generalized mapping as expressed in Eq. (7.20) makes it impossible to evaluate the above convergence criteria for the mapping in the general case, but once one introduces a specific functional form for φ then the problem is more amenable to solution. We demonstrate this for a particular example. From this section onwards, we will use the classical function representation for signals except where indicated.

Introducing a specific form $\varphi(x) = e^{(x \cdot x') * h_f}$, it is seen that the GMF in Eq. (7.17) reduces to

$$x_{p+1} = x_p + \lambda(u - x_p * h). \tag{7.29}$$

The above equation is the pioneering van Cittert restoration method[14] proposed by van Cittert in 1931.

Taking the Fourier transform of both sides, we now rewrite this equation in the transform domain as follows:

$$X_{p+1} = X_p + \lambda(U - H \cdot X_p), \tag{7.30}$$

where upper case letters have been used to denote the transforms of the signals, from the time or spatial domain.

Since the Fourier transform preserves the relevant properties of the space $\mathcal{S}$, the condition for convergence remains the same as given in Eq. (7.26). Continuing with our analysis, notice that the first variation of the right-hand side of Eq. (7.30), with respect to X is

$$1 - \lambda H(w) \tag{7.31}$$

for each w, where $\mathbf{w}$ is the discrete frequency vector with k elements. Therefore, from Eq. (7.26) we require that

$$|1 - \lambda H(w)| < 1 \quad, \forall w. \tag{7.32}$$

Thus for convergence of this algorithm at each iteration λ has to be constrained so that

$$-1 < 1 - \lambda H(w) < 1. \tag{7.33}$$

Rewriting Eq. (7.33) this we see that the condition for convergence can be expressed as shown below:

$$0 < \lambda H(w) < 2, \tag{7.34}$$

which is the well known condition for convergence of the van Cittert algorithm with a relaxation parameter, described by Jansson.[14]

The nature of the present problem is such that the original data is bandlimited, and the system function is usually low-pass with a maximum at $w = 0$. When this is true, the above condition on λ is most restrictive when $w = 0$. We can then guarantee that λ satisfies Eq. (7.32), if $H(w)$ is replaced by $H(0)$. In the implementation of deconvolution algorithms, it is typical practice to normalize the system response function so that the energy in the original image is conserved. The following equation expresses this mathematically:

$$\sum_{i=1}^{K} h(i) = 1. \tag{7.35}$$

This implies that $H(0) = \frac{1}{2\pi}$. With this information and by the reasoning given in the preceding paragraph, the condition for convergence may be written as a simple condition on λ:

$$0 < \lambda < 4\pi. \tag{7.36}$$

The condition on h for convergence of the van Cittert algorithm when $\lambda = 1$ is well known and has been discussed earlier by Hill and Ioup.[15]

7.2.4 Incorporating hard constraints

We have discussed the use of a constraint operator $\mathbf{C}^3$ to represent the use of hard constraints, such as positivity and band limitation. If such an operator is to be used, then the composition of the two operators $\mathcal{T}, \mathbf{C}$ (or mappings) has to be a contraction. This does not change the analysis, since $\mathbf{C}$ is known to be a nonexpansive map.[16] As a result, the composite mapping is a contraction if $\mathcal{T}$ is a contraction.

7.2.5 Making λ adaptive

Generally, it is desirable that the step size or smoothing parameter λ be adaptively updated at each iteration. It is necessary that this process be implemented in such a fashion that the rate of convergence is not affected negatively. Obviously, such an *optimum* computation of λ is intimately related to the particular form of the mapping being investigated. For some specific cases, Achour[11] has presented a method of determining λ in a manner consistent with the above discussion. In this technique, λ is calculated at each iteration by solving a quadratic function in λ. If the equation so derived fails to have a real root, then a steepest descent-type value is chosen for λ by setting the first variation of the quadratic function equal to zero. The quadratic function itself is obtained by using the noise-variance constraint presented in Eq. (7.14). For more details see Ref. 11.

7.3 Conclusion

Functionals arising in information theory may also be used to regularize ill-posed problems encountered in signal or image processing. The MI functional, measuring the dependence between two random variables, is particularly useful in constructing "worst-case" solutions to such problems. Key to this approach is the use of a GMF, an invertible map between the signal or image space and the space of PDFs. While such a mapping precludes a strict probabilistic interpretation, it enables a fast and generally convergent iterative algorithm, which may be used to tackle various signal-processing tasks.

References

1. B. Frieden, "Restoring with maximum likelihood and maximum entropy," *JOSA* **62**, 511–518 (1972).

2. J. Marsden and M. Hoffman, *Elementary Classical Analysis*, WH Freeman, New York, NY (1993).

3. A. Katsaggelos, J. Biemond, R. Schafer, and R. Mersereau, "Λ regularized iterative image restoration algorithm," *IEEE Transactions on Signal Processing* **39**(4), 914–929 (1991).

4. J. Burg, *Maximum Entropy Spectral Analysis*, Stanford University, Stanford, CA (1975).

5. J. Noonan, N. Tzannes, and T. Costello, "On the inverse problem of entropy maximizations," *IEEE Transactions on Information Theory* **22**(1), 120–123 (1976).

6. N. Tzannes and J. Noonan, "The mutual information principle and applications," *Information and Control* **22**(1), 1–12 (1973).

7. J. Noonan and J. Marcus, "Minimum mutual information in image restoration," *Kybernetes* **19**(6), 34–41 (1990).

8. J. Noonan and J. Marcus, "An approach to stochastic systems modelling," *Kybernetes* **15**(4), 225–229 (1986).

9. J. Noonan and B. Achour, "A hybrid mip algorithm for signal restoration," in *IASTED Intl. Conf. on Adaptive Control and Signal Processing*, 93–97, IASTED, (NY) (1990).

10. J. Noonan and B. Achour, "Two new robust nonlinear signal restoration algorithms," *Digital Signal Processing* **2**(1), 39–43 (1992).

11. B. Achour, *Regularization Theory in Signal Restoration – An Information Theoretic Approach*. PhD thesis, Tufts University (1991).

12. J. Noonan, P. Natarajan, and B. Achour, "A class of iterative signal restoration algorithms," in *1995 IEEE International Symposium on Information Theory, 1995. Proceedings.*, 115 (1995).

13. A. Goldstein, *Constructive Real Analysis*, Harper & Row, New York, NY (1967).

14. P. Jansson, *Deconvolution: With Applications in Spectroscopy*, Academic Press, San Diego, CA (1995).

15. N. Hill and G. Ioup, "Convergence of the van Cittert iterative method of deconvolution," *JOSA* **66**(5), 487–489 (1976).

16. D. Youla and H. Webb, "Image reconstruction by the method of convex projections: Part 1," *IEEE Transactions on Medical Imaging* **1**(2), 81–94 (1982).

Chapter 8
Special Cases of the GMF

8.1 Introduction

In Chapter 3, various methods for solving the inverse problem were visited. Among those, the iterative methods have gained the greatest popularity, especially since the advent of digital computers. Conventional methods present major difficulties in implementation. Typically, a matrix needs to be inverted. Large matrices call for storage and computer time in order to be handled. Singular and ill-conditioned matrices make the problem very sensitive to noise and therefore unstable or impossible to solve. Iterative methods are attractive; they avoid the matrix inversion and approach the solution by successive approximations. Also, iterative methods have the advantage of allowing the scientist to control the false information in the estimate by interaction with the solution as it evolves. This may be achieved either automatically in the algorithm or by the exercise of human judgement.

Early work in the deconvolution by successive approximations resulted in various linear methods. These methods have relatively poor performance, especially with bandlimited data. However, when fast computation is desired, linear methods proved useful.

Usually, in signal recovery, some form of knowledge about the system is available, such as the statistics of the corrupting noise or some information about the anticipated solution. Therefore, a good method for the inverse problem is one that uses all of the available information about the system. This leads us to the so-called modern-constrained method or, simply, nonlinear methods. These methods are inherently more robust than the linear ones because they must find a solution that is consistent with both data and physical reality. Sometimes, physical reality is referred to as the prior information about the system. The prior information is imposed as constraints on the mathematical model that describes the method.

In the past, linear methods dominated deconvolution research because of their simpler analysis and shorter computations relative to nonlinear methods. Computational speed is also an advantage of linear methods. However, the capability of these methods to restore signals is very limited. Noisy data, for example, often give rise to physically untrue information, for instance, negative pixel values in image processing. Therefore, one is driven to develop restoration methods that will allow the incorporation of physical requirements. This prior information may be introduced in the analysis process in the form of constraints. This goal is achieved by nonlinear methods. The nontrivial mathematics and the high computational cost involved discouraged many researchers from looking closer at these methods. However, the advent of digital computers and the desperate need for better restoration techniques made the future of nonlinear methods certain. Fortunately, the advances in computer hardware made computation time less of a problem and nonlinear techniques more attractive.

As mentioned earlier, it is desirable to be able to incorporate any kind of *a priori* information in our formulation of the problem. We often use the word *constraints* to characterize this available information. A popular constraint in the fields of image processing and spectroscopy is positivity. That is, the expected solution is not allowed to take on negative values. A more general form of this constraint is to put bounds on the solution. Other constraints might represent some known statistics of the expected solution or the contaminating noise. Finite support of the anticipated solution is also a common constraint, especially in the frequency domain where this is known as the finite bandwidth constraint.

The GMF, in the form of Eq. (7.20), represents a family of iterative signal-restoration algorithms. It has already been seen that the van Cittert method is one specific form of the GMF; other choices of φ yield different forms of the GMF. In this chapter, we will focus on one specific form of the mapping, and show that three popular restoration algorithms are either directly identifiable to this form or are simple variants. All three algorithms were originally derived from other considerations of optimality. As a prelude to presenting the variants, we state the GMF in algorithmic form, which serves as an appropriate background for the subsequent discussion.

$$x_0 = \text{prior estimate, e.g., white noise or the data itself,} \tag{8.1}$$

$$x_{p+1} = \varphi^{-1}\left(\varphi(x_p)exp\left(\lambda x_p'\left(h_f \cdot [y - h * x_p]\right)\right)\right). \tag{8.2}$$

In the following, we derive a particular form of the GMF, which is used in the rest of the chapter. In Eq. (7.17), if φ is chosen such that

$$\varphi(x) = e^{xx'}, \tag{8.3}$$

where

$$x' = \frac{\partial x}{\partial \varphi}, \tag{8.4}$$

then, the GMF reduces to the following specific form:

$$x_{p+1} = x_p + \lambda(y - h * x_p) * h_f. \tag{8.5}$$

Incorporating the constraint $\mathbf{C}$ into the above equation explicitly, we obtain

$$x_{p+1} = \mathbf{C}x_p + \lambda(y - h * \mathbf{C}x_p) * h_f. \tag{8.6}$$

8.2 Gradient-based Technique

For the restoration of synthetic aperture sonar data, Ochieng-Ogolla[1] and others have proposed an algorithm based on an optimal gradient technique for solving a linear system of equations. This approach is based on the fact that gradient techniques that minimize the quadratic function

$$f(\mathbf{u}) = \frac{1}{2}\mathbf{u}^T \mathbf{H}\mathbf{u} - \mathbf{v}^T \mathbf{u} + c_1 \tag{8.7}$$

solve a linear system similar to the degradation model* in Eq. (2.11), i.e.,

$$\mathbf{v} = \mathbf{H}\mathbf{u}. \tag{8.8}$$

The following algorithm is then derived using principles of optimization theory and by modifying the basic iteration in Eq. (3.14)

$$x_{p+1} = \mathbf{C}x_p + \alpha_k h \otimes (y - h * \mathbf{C}x_p). \tag{8.9}$$

In the above equation, $\otimes$ denotes the correlation of two signals. It is well known that correlation of two signals is simply convolution of one signal with a flipped version, i.e., time reversed in the case of temporal signals, of the other signal. In the second term on the right-hand side of Eq. (8.9) we may replace the correlation with convolution and rewrite the equation as

$$x_{p+1} = \mathbf{C}x_p + \alpha_k h_f * (y - h * \mathbf{C}x_p). \tag{8.10}$$

A comparison of Eq. (8.10) and Eq. (8.6) shows the identical nature of the two restoration algorithms. The method presented in Ref. 1 is thus a particular case of the GMF.

8.3 Least Squares as a Special Case

Angel and Jain[2] presented the following least-squares estimate (LSE) algorithm for signal restoration in 1977. The algorithm is given by the iteration

$$\hat{x}^+(i,j) = \hat{x}^-(i,j) +$$
$$\gamma \sum_{l,m=1}^{I} \sum^{J} h(i,j;l,m) \left[y(l,m) - \tilde{b}(l,m) \right], \tag{8.11}$$

where

$$\tilde{b}(i,j) = \sum_{l,m=1}^{I} \sum^{J} h(i,j;l,m) \hat{z}(l,m). \tag{8.12}$$

---

*The difference being that in the present model, the effect of noise is not incorporated.

h is the PSF of the blurring system. The equation above is for the general space-variant case. For the space-invariant case, which is the case of interest here, the summations in the above equations reduce to convolution relationships, .i.e.,

$$\hat{x}^+ = \hat{x}^- + \gamma h * \left(y - \tilde{b} \right),$$ (8.13)

where

$$\tilde{b} = h * \hat{x}.$$ (8.14)

In Eq. (8.5), let $h_f = h$. We then see that Eqs. (8.5) and (8.13) are the same. In the discussion following the derivation of the GMF, it was mentioned that any change to the structure of the GMF *could*, but not necessarily, lead to suboptimal performance. If $h_f = h$, this is an example of one variant that is suboptimal in the MIM sense but optimal in the mean-square sense. Also, the PSF, or the distorting function, h is almost always a symmetric function. Therefore setting $h_f = h$ does not lead to any deviation from optimal performance for those cases.

8.4 Maximum Likelihood as a Special Case

The following ML estimator in the presence of additive white Gaussian noise (AWGN) was recently suggested by Namazi and Fan.[3] This algorithm was derived using the same distortion model as in Eqs. (2.10) and (2.11) and is expressed as

$$\hat{x}^+ (\imath, \jmath) = \hat{x}^- (\imath, \jmath) +$$
$$\gamma \sum_{l,m}^{I} \sum_{=1}^{J} COV_{xb} (\imath, \jmath; l, m) \left[y(l, m) - \tilde{b}(l, m) \right].$$ (8.15)

Here, $+$ denotes the new estimate and $-$ the old estimate. COV_{xb} is the cross-covariance function between x and b.

Once again, the equations above give the solution for a space-varying PSF. As mentioned by the authors Namazi and Fan, in the case where the PSF is space invariant and $x(\imath, \jmath)$ is a sample function from a stationary random field, the summations in Eqs. (8.15) and (8.14) reduce to convolution relationships, i.e.,

$$\hat{x}^+ = \hat{x}^- + \gamma COV_{xb} * \left[y - \tilde{b} \right],$$ (8.16)

and $\tilde{b}$ is as in Eq. (8.14) in the preceding section. Using the same specific form of the GMF as in Eq. (8.5) we can see that the above iteration in Eq. (8.16) is a special case of the GMF with an appropriate choice of $h_f = COV_{xb}$.

A comparison of Eqs. (8.16) and (8.5) shows that for particular choices of h and h_f, the ML estimator is indeed a special case of the GMF.

8.5 Convergence of the Particular Variants

We now discuss the convergence of the iteration given by Eq. (8.5). Using the convergence criteria derived in Chapter 7, we subsequently develop a specific condition tailored to the mapping of Eq. (8.3). In this section we use matrix notation to compare the convergence condition presented here with that in the literature on the subject.

Let φ be chosen as follows:

$$\varphi\left(\mathbf{x}\right) = e^{\mathbf{xx}'}, \tag{8.17}$$

then the GMF reduces to the following specific form:

$$\mathbf{x}_{p+1} = \mathbf{x}_p + \lambda \mathbf{H}^T \left(\mathbf{y} - \mathbf{H}\mathbf{x}_p\right). \tag{8.18}$$

In this case the mapping $\mathcal{T}$ is given by

$$\mathcal{T}\mathbf{x}_p = \mathbf{x}_p + \lambda \mathbf{H}^T \left(\mathbf{y} - \mathbf{H}\mathbf{x}_p\right). \tag{8.19}$$

The first variation of this with respect to $\mathbf{x}_p$ is

$$1 - \lambda \mathbf{H}^T, \tag{8.20}$$

and for $\mathcal{T}$ to be a contraction we require that

$$\overset{\max}{\underset{\mathbf{x},i,j}{}} \left|1 - \lambda \mathbf{H}^T \mathbf{H}\right| < 1. \tag{8.21}$$

Obviously, the above equation is independent of $\mathbf{x}$. Simplifying this equation, the following inequality is obtained

$$0 < \lambda \overset{\max}{\underset{i,j}{}} \left\{\mathbf{H}^T \mathbf{H}\right\} < 2, \tag{8.22}$$

where i, j are the matrix element indices, and finally we have the following condition on λ for $\mathcal{T}$ to be a contraction:

$$0 < \lambda < \frac{2}{\overset{\max}{\underset{i,j}{}} \left\{\mathbf{H}^T \mathbf{H}\right\}}. \tag{8.23}$$

For related convergence analysis, see Refs. 4 and 1. There the convergence condition is derived for a slightly modified case, but the basic bounds on λ can be seen to be the same.

8.6 Conclusion

The GMF enables a framework by which the MIM algorithm may be viewed as a *generalization* of several well-known algorithms, including the least-squared error algorithm and the ML algorithm. Indeed, so long as the GMF is invertible, almost any convex regularization functional can be derived from the MIM formulation. As such, properties such as convergence and stability need only be verified once for the MIM algorithm and not for each individual variant, provided a suitable mapping function exists.

References

1. E. Ochieng-Ogolla, S. Fischer, A. Wasiljeff, and P. de Heering, "Iterative restoration algorithm for real-time processing of broadband synthetic aperture sonar data," *Real-Time Imaging* **1**(1), 19–31 (1995).

2. E. S. Angel and A. Jain, "Restoration of images degraded by spatially varying pointspread functions by a conjugate gradient method," *Applied Optics* **17**(14), 2186–2190 (1978).

3. N. Namazi and C. Fan, "A new iterative algorithm for image restoration based on the maximum likelihood principle," *International Journal of Modeling and Simulation* **14**, 6–11 (1994).

4. A. Katsaggelos, J. Biemond, R. Schafer, and R. Mersereau, "A regularized iterative image restoration algorithm," *IEEE Transactions on Signal Processing* **39**(4), 914–929 (1991).

Chapter 9
Applications

9.1 Introduction

We conclude this monograph by presenting applications of the developed algorithm to common problems in signal processing. We have demonstrated that the MIM algorithm is well suited to problems where the solution is sparse in some domain. To wit, the MIM algorithm yields good solutions to deconvolution problems that are "well structured" and "peaky." Below, we give two applications that leverage this property. First, we show how the MIM algorithm may be used to obtain spectral density estimates with sharper frequency resolution. We then conclude with an extended treatment of an application of the MIM algorithm to blind text restoration, exploiting sparse representations of Roman alphabets.

9.2 Spectral Estimation

Spectral-density estimates are often obtained via the fast Fourier Transform (FFT). Typically, such FFT-based estimates are of low resolution; time-domain windows used to suppress unreliable autocorrelation coefficients also act to blur the spectral estimate. To rectify this loss of resolution, we require *robust* deconvolution techniques to sharpen the image while still maintaining SNR improvements gained via windowing.

Estimating a unique power spectral density given a truncated version of the autocorrelation function is ill defined, as there may exist an infinite number of spectra consistent with the given data. Well-known methods of spectral estimation, including those exploiting the Wiener–Khintchine equality, typically assume that unknown values of the autocorrelation function are uniformly zero. In an information-theoretic setting such an approach is suspect; the assumption that unknown values are zero infers information about the process that may not be substantiated by the data. In this section, we present a more principled approach to estimation in this setting derived via the mutual information principle (MIP).

9.2.1 Problem statement and model

Assume that we are given an autocorrelation sequence $R_{zz}[m]$ of a stationary discrete time series sequence $z[n]$. Often, such sequences may contain errors imparted

73

by data windows, or the time series $z[n]$ itself may be corrupted by noise. Let $R_{yy}[m]$ denote the true autocorrelation sequence for which we wish to compute the spectral estimate $S_{yy}[k]$. We may characterize the noise or uncertainty present in $R_{zz}[m]$ via the constraint equation:

$$\frac{1}{M} \sum_{m=1}^{M} (R_{zz}[m] - R_{yy}[m])^2 = \sigma^2, \tag{9.1}$$

where σ^2 is the power of the modeled uncertainty process, assumed known. To estimate the spectral density $S_{yy}[k]$, we frame the estimation as an inverse problem. To wit, we may write:

$$\mathfrak{F}(S_{yy}[k]) + N[m] = R_{zz}[m], \tag{9.2}$$

where $\mathfrak{F}(\cdot)$ denotes the Fourier operator, and N is the modeled uncertainty process.

9.2.2 Approach

The form of Eq. (9.2) suggests using standard regression to compute an estimate of $S_{yy}[m]$. However, standard regression techniques in solving this inverse problem may favor candidate solutions with high degrees of smoothness, sacrificing resolvability among different spectral peaks. On the other hand, we have seen that applications of the MIP produce much sharper estimates. Accordingly, we now show how the MIP may be applied to yield the *minimum relative entropy (MRE)* spectral estimate.[1]

To regularize Eq. (9.1), we begin with the relative entropy functional:

$$\sum_{k=1}^{K} S_{yy}[k] \log \frac{S_{yy}[k]}{P[k]}, \tag{9.3}$$

where $S_{yy}[k]$ is an estimate of the MRE spectrum and $P[k]$ is the prior guess of $S_{yy}[k]$. Thus, the constraint equation subject to which Eq. (9.3) is minimized is given by Eq. (9.1).

To solve this minimization problem, we employ the method of Lagrangian multipliers from the calculus of variations. We may write

$$\frac{\partial}{\partial S_{yy}[k]} \left(\sum_{k=1}^{K} S_{yy}[k] \log \frac{S_{yy}[k]}{P[k]} + \lambda \frac{1}{M} \sum_{m=1}^{M} (R_{zz}[m] - R_{yy}[m])^2 \right) = 0 \quad \forall k. \tag{9.4}$$

We implicitly define $H[m]$ via the relation:

$$R_{yy}[m] = \sum_{k} S_{yy}[k] H[m - k]. \tag{9.5}$$

Then, Eq. (9.4) yields

$$1 + \log \frac{S_{yy}[k]}{P[k]} - \frac{2\lambda}{M} \sum_{m=1}^{M} (R_{zz}[m] - R_{yy}[m])H[m-k] = 0 \quad \forall k. \qquad (9.6)$$

We now introduce the iterative nature of the method. Let $S_{yy}^{(p)}[k]$ denote the pth estimate of the enhanced spectrum $S_{yy}[k]$, and let $P[k]$ be the first estimate of $S_{yy}[k]$. From this an iteration on $R_{yy}[m]$ follows as

$$R_{yy}^{(p)}[m] = \sum_{k} S_{yy}^{(p)}[k]e^{j2\pi mk}. \qquad (9.7)$$

As $S_{yy}[k]$ is real and hermitian, we may equivalently write

$$R_{yy}^{(p)}[m] = \sum_{k} S_{yy}^{(p)}[k]\cos(2\pi mk). \qquad (9.8)$$

The introduction of the iterative estimates of S_{yy} and R_{yy} allows Eq. (9.3) to be rewritten slightly differently as

$$1 + \log \frac{S_{yy}^{(p)}[k]}{S_{yy}^{(p-1)}[k]} - \frac{2\lambda}{M} \sum_{m=1}^{M} (R_{zz}[m] - R_{yy}^{(p-1)}[m])\cos(2\pi mk) = 0 \quad \forall k, \quad (9.9)$$

which in turn gives:

$$S_{yy}^{(p)}[k] = S_{yy}^{(p-1)}[k]e^{-1+\nu \sum_{m=1}^{M}(R_{zz}[m]-R_{yy}^{(p-1)}[m])\cos(2\pi mk)}, \qquad (9.10)$$

where $\nu = \frac{2\lambda}{M}$ is the Lagrangian multiplier, which may be found at each step by solving

$$\frac{1}{M} \sum_{m=1}^{M} (R_{zz}[m] - R_{yy}^{(p)}[m])^2 = \sigma^2. \qquad (9.11)$$

We have therefore derived an iterative restoration algorithm from an application of the MRE principle (or equivalently the MIP). The existence and uniqueness of the minimum of the constrained MRE functional is guaranteed by the positivity of its second derivative.

9.2.3 Applications and examples

The most likely application of the method presented above is to augment a spectral estimate obtained using noisy autocorrelation lags. That is, assuming that N noisy autocorrelation lags are available, we apply general spectral estimation methods treating the noisy autocorrelation lags as exact. This provides a spectral estimate that can be used as the starting point for the algorithm. The constraint equation is used iteratively; each new posterior MRE estimate serves as a prior guess for the

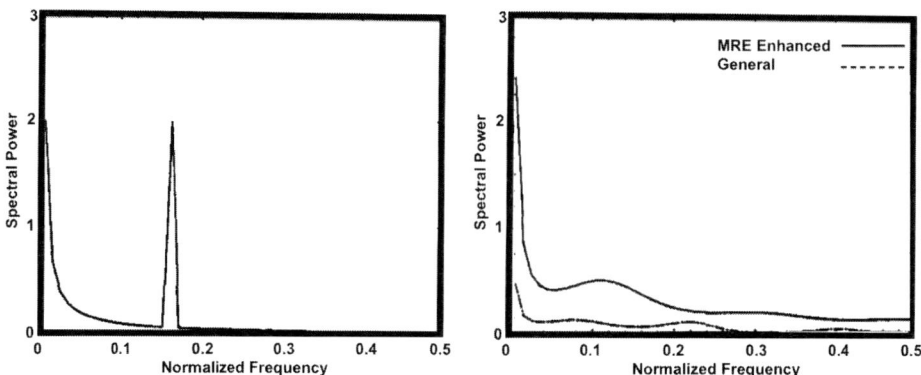

Figure 9.1 True (left panel) and enhanced spectra (right panel) for the first example.

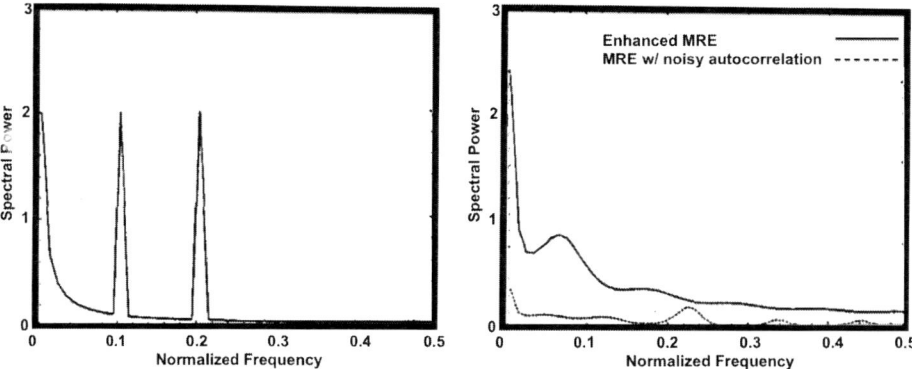

Figure 9.2 True (left panel) and enhanced spectra (right panel) for the second example.

next step. Of course, at each iteration it is also necessary to evaluate the most recent estimate of $R_{yy}^{(p)}$ provided by Eq. (9.8).

Two examples are presented in which the autocorrelation function is corrupted by AWGN with zero mean and variance such that a 3-dB signal-to-noise ratio (SNR) results.

9.2.3.1 Example 1

In the first example, the true spectrum is as shown in the Fig. 9.1. The autocorrelation function (lags are assumed at unity spacing) is obtained exactly (using a Fourier transform). Only six of these are retained, and AWGN—with an SNR of 3dB—is added to each of them. The general spectral estimate that results by treating the noisy estimates as exact is shown in the right panel, along with the MRE-enhanced estimate that uses knowledge of the noise variance as a constraint. The improvement provided by the new method is obvious.

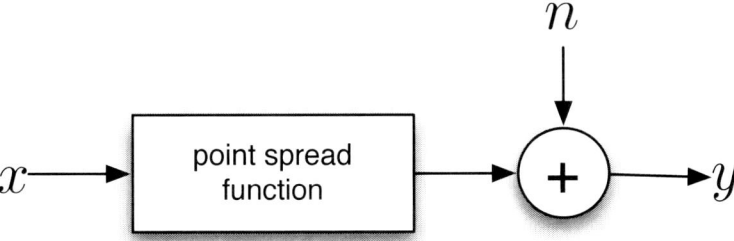

Figure 9.3 Degradation model of noisy observed images.

9.2.3.2 Example 2

As a second example, we consider the true spectrum shown in Fig. 9.2. As in the first example, the exact autocorrelation function is obtained via the FFT. For the application, ten noise-corrupted (SNR = 3 dB) autocorrelation function lags are all that is assumed known. The general and enhanced spectral estimates are shown in the right panel. Again, the improvement in the estimate is obvious.

9.3 Image Restoration

We now study the performance of the MIM algorithm in the estimation of two dimensional signals. A natural and standard setting for such an exploration is the restoration of blurred or noisy images. Typically, the characteristics of the image degradation process are assumed known under some model and image restoration consists of inverting this degradation.

9.3.1 Model

Figure 9.3 illustrates the standard linear model assumed for such problems: the blurring of the clean input image x is modeled by some assumed PSF. This output is then further corrupted by AWGN to produce the final observation. Below, we give further justification for two components of the model: the form of the PSF and the assumption of normality of the noise statistics.

Defocus refers to any departure along the optical axis from the point where the light rays of a captured object converge (focusing plane). As Fig. 9.4 shows, this creates a circular blur in the acquired image for each point source being captured. This blur is specified via the parameter[2]

$$\sigma = \rho r v \left(\frac{1}{f} - \frac{1}{v} - \frac{1}{u} \right),\qquad(9.12)$$

where f is the focal length, u is the distance from the point source to the lens, v is the distance from the lens to the image plane of the detector, r is the radius of the

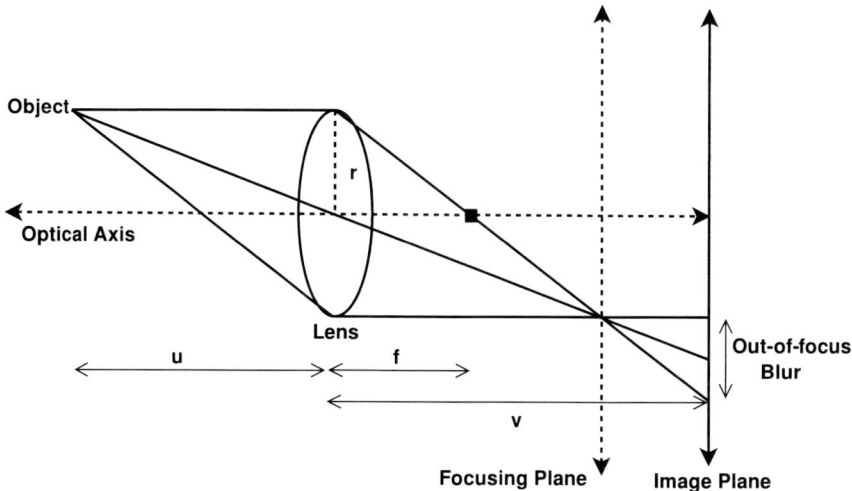

Figure 9.4 Defocused image formation.

lens, and ρ is a constant depending on the optics of the camera. The defocus blur is usually modeled by either a uniform intensity circle of radius σ or a Gaussian spread function of variance σ^2. We consider only the Gaussian blur case so that the form of the PSF is given by

$$h(i,j) = \frac{1}{2\pi\sigma^2} e^{-\frac{i^2+j^2}{2\sigma^2}}. \tag{9.13}$$

Noise is often modeled as a Gaussian process in image-processing problems. Such an assumption is often accurately derided as invalid in the context of image data; image noise is known to more faithfully obey a Laplacian distribution. However, the Gaussian assumption is nonetheless useful for one important reason. As established in Ref. 3, the Gaussian assumption leads to the worst Cramer–Rao bound for any given estimator. Thus, any estimator attaining or coming close to the Cramer–Rao bound can be interpreted as minimax optimal.

9.3.2 Approach

The extension of the MIM estimation algorithm to the two-dimensional case is straightforward. Recalling Eq. (7.20), we have:

$$\mathbf{x}_{p+1} = \varphi^{-1}\left(\varphi(\mathbf{x}_p)\exp\left(\lambda\mathbf{x}_p'\left(\mathbf{H}^T\cdot[\mathbf{y}-\mathbf{H}\cdot\mathbf{x}_p]\right)\right)\right). \tag{9.14}$$

Here, the vector $\mathbf{y}$ represents a stacked ordering of the pixel values in the observed image y, $\mathbf{H}$ the convolutional matrix describing the two-dimensional blurring of the PSF h, and $\mathbf{x}$ the stacked estimate. The function φ scales the image, so that the sum of all pixel values in its argument sums to one. Finally, λ, the adaptive step size, may be calculated by the usual Lagrangian methods.

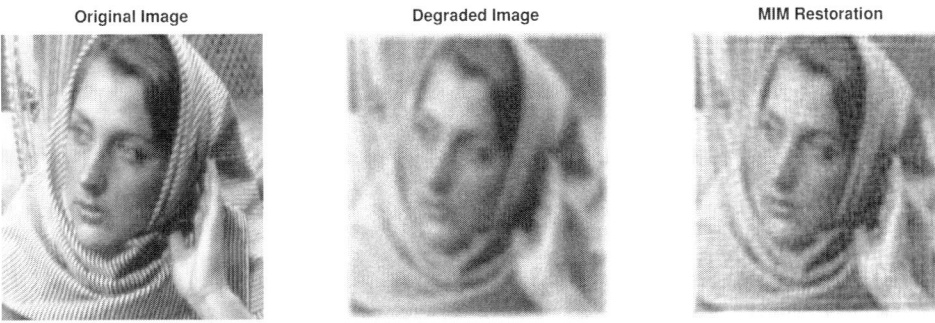

Figure 9.5 Image restoration of the "Barbara" image using MIM.

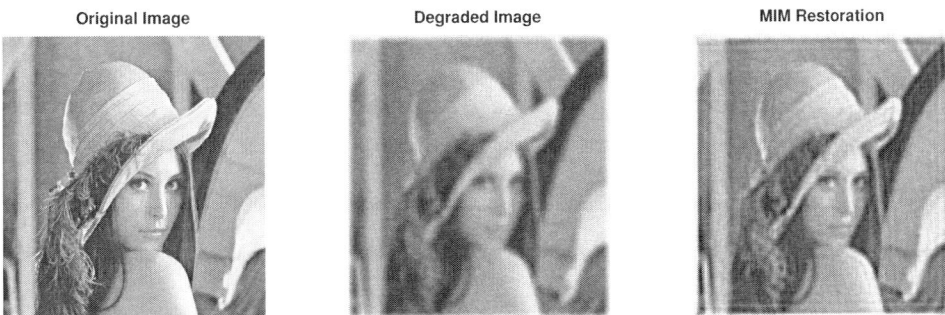

Figure 9.6 Image restoration of the "Lena" image using MIM.

9.3.3 Results

Figures 9.5 and 9.6 show two examples of applications of the MIM algorithm for image restoration. In Fig. 9.5, the original 240×240 image is blurred by a Gaussian kernel of standard deviation 8 pixels and corrupted by noise to yield a 20-dB output SNR. The restored image is shown on the right panel. Observe that many of the high-frequency features (e.g., the stripes in the headscarf) absent in the blurred image have been recovered to some extent in the restoration. Similar results are seen in Fig. 9.6 especially in the fine texture information recovered in the feathers of the boa.

9.4 Text Image Restoration

With the recent prevalence of low-resolution charge-coupled devices (CCDs), such as those found in cell phones and security cameras, there has been a renewed interest in the blind restoration problem. Where text images are concerned, the problem becomes one of *resolution expansion*. An efficient solution to this problem is required for many problems in homeland security. For example, accurate restorations would allow low-resolution surveillance cameras to be used to identify and track the license plates of cars leaving the scene of a crime. In the context of surveillance or espionage, low-resolution cameras, such as those found in many mobile phones, could be used to adequately capture the contents of a document.

In these problems, the goal is to expand the resolution of a given observed low-resolution image and to negate the distorting effects of defocus and the CCD sensor PSF. However, for any given image there are an uncountable number of super-resolution images that fit the observed data. Thus, the resolution-expansion problem falls into the general class of ill-posed inverse problems.

In this section, we consider solutions to this problem using the MIM algorithm. We will exploit the fact that the MIM algorithm gives nonnegative and somewhat well-structured solutions, making it a good choice for text restoration. While the algorithm demonstrates reasonable performance in the spatial domain, we will consider solutions in a domain more suited to exploiting the nature of the MIM solutions.

9.4.1 Problem statement and model

Given a single low-resolution defocused image of text from noncursive script (e.g., Roman, Cyrillic, Kanji, etc.), we wish to recover the original input image. This blind single-input single-output (SISO) deconvolution problem may be modeled by two separate processes, defocus and low resolution. We formulate the models for each below.

9.4.2 Defocus and low resolution

We have already discussed the blurring model in the previous section. In the true blind case, the information required to calculate σ in Eq.(9.12) is unknown. Thus, we will assume that σ is unknown in Eq. (9.13).

Having reached the image plane, the light rays emitted from the point source shown in Fig. 9.4 are captured by an array of sensors. As there is a limit to the number of sensors available to capture the image data, the acquired image must necessarily lose some information emitted from the source. To model this process,[4] we define x as the original perfect resolution scene, our text image, captured at some tractable resolution. At the image plane, the high-resolution blurred image is given by $y = h * x + n$. If the sensor array is assumed to degrade this desired high resolution by some factor q, then the observed image will be given by

$$y_d(i,j) = \frac{1}{q^2} \sum_{k=qi}^{(q+1)i-1} \sum_{l=qj}^{(q+1)j-1} y(k,l) \qquad i,j = 1 \ldots N. \qquad (9.15)$$

Thus, the low-resolution detection process can be thought of as the action of each sensor in the array averaging q^2 pixels from the desired high-resolution scene. Taking Y_d as the $N^2 \times 1$ lexicographically ordered vector containing the pixel values from the $N \times N$ matrix y_d, X as the lexicographically ordered high-resolution

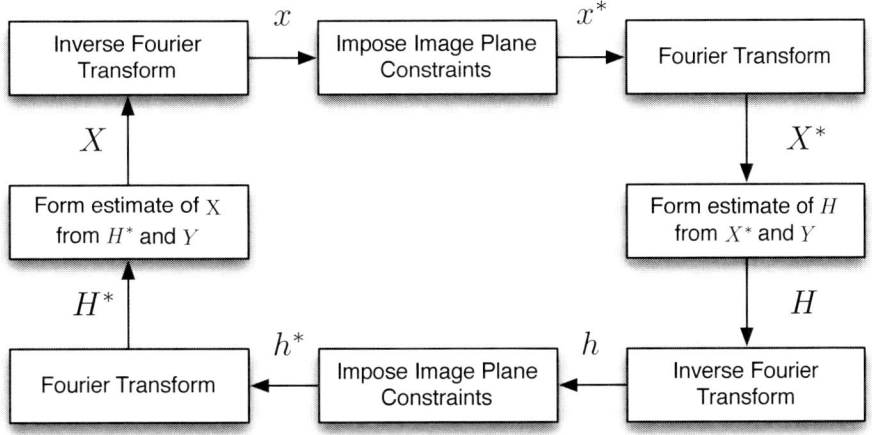

Figure 9.7 Flow diagram of IBD

scene, D as the decimation matrix, and H as the blurring matrix, we can write Eq. (9.15) as[5]

$$Y_d = DHX + n, \tag{9.16}$$

where n is now the lexicographically ordered decimated noise process.

9.4.3 Approach

Under the model assumptions in the previous section, we propose an approach to the super-resolution problem motivated by the iterative blind deconvolution framework first proposed in Ref. 6. In Ref. 6, the iterative blind deconvolution (IBD) method is introduced as an alternating minimization blind deconvolution algorithm that alternates between constraints in the Fourier domain and the image domain. For the observed image $y = h * x + n$, where x denotes the image, h denotes the blurring kernel, and n denotes added Gaussian noise, the technique is illustrated schematically in Fig. 9.7.

Our approach here is to modify the IBD procedure to exploit certain properties of a deconvolution algorithm proposed by Noonan et al. in Refs. 7, 8, and 9, among other works. However, before discussing the details of this approach, we first introduce its major components.

9.4.3.1 The Radon transform

One of the problems with SISO blind deconvolution techniques, especially in comparison to multiple-input single-output (MISO) deconvolution problems, is the relative sparsity of available information with which to regularize the two underlying ill-posed problems contained in our degradation model [Eq. (9.16)]. Ideally, since the image domain offers little *a priori* information, we would like to find some

Image Domain   Radon Domain   Image Domain   Radon Domain

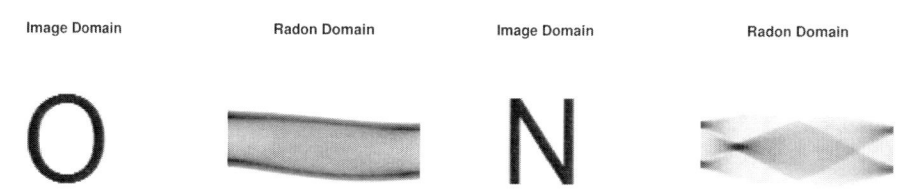

Figure 9.8 Radon Transform of the letters o and n demonstrating the "peakiness" of the transform on Roman letters.

transformation of the image that gives each admissible solution, i.e., the space of noncursive letter images, a common characteristic. Based on some empirical images, we propose that the Radon transform is such a method. Recall that the Radon transform is given by

$$x_r(\rho, \theta) = \int \int x(\alpha, \beta)\delta(\rho - \alpha\cos\theta - \beta\sin\theta)d\alpha d\beta \qquad (9.17)$$

in the continuous case, and

$$x_r(\rho, \theta) = \sum_{\alpha}\sum_{\beta} x(\alpha, \beta)\delta(\rho - \alpha\cos\theta - \beta\sin\theta)$$

$$\rho = 1, \ldots, N \quad \theta = 0, \ldots, 179 \quad (9.18)$$

for the discrete case. The Radon transform takes lines in the spatial domain and maps them to points in the Radon domain. We are thus exploiting the fact that the noncursive alphabet consists of many letters that can be modeled as a union of straight lines. As Fig. 9.8 demonstrates, the Radon is a sufficiently peaky transform for such letters. For curved letters, such as the letter "o," the Radon maintains this peakiness, albeit only in the vertical plane. However, for reasons we shall see later, this is not a very big problem.

One potential drawback of working in the Radon domain is that the transform of a whole word is not necessarily compacted in nature. We must thus get around this by partitioning the text document into letter-sized segments and then go about restoring each of these segments individually.

A key result relating to the Radon transform is due to the projection-slice theorem. It implies that the Radon transform of the two-dimensional convolution of two functions is equal to the one-dimensional convolution of their Radon transforms. Thus, we have that for a fixed θ_0 and for some unknown H:

$$R_Y(\rho, \theta_0) = R_X(\rho, \theta_0) * R_H(\rho, \theta_0), \qquad (9.19)$$

where $R_X(\rho, \theta_0)$ denotes the Radon transform vector of x at angle θ_0. If the channel is rotationally invariant, the equation above becomes

$$R_H(\rho, \theta_i) = R_H(\rho, \theta_j) \qquad \forall\theta_i, \theta_j \in \Theta. \qquad (9.20)$$

The advantages of the structure present in the Radon transform of letter images are two-fold. First, consider the deaveraging problem. In the image domain, the deaveraging process necessarily involves some sort of interpolation. While there are sophisticated interpolation schemes for deaveraging text images (see Ref. 10), the fastest and simplest approach, linear interpolation, offers little in the way of resolution improvement. However, in the Radon domain, applying linear interpolation produces much better results since the text images will have an underlying exploitable structure. Second, as we shall see in the next section, the low-entropy characteristic of Radon-transformed text images can also be used when regularizing the ill-posed deconvolution problem arising from undoing the effects of the PSF.

9.4.3.2 MIM algorithm

As we have already noted, the model given in Eq. (9.16) consists of an averaging operation and a blurring operation. For our restoration approach, we seek to undo these distortion effects separately. However, we will also leverage characteristics of each solution in solving the dual problem. The deaveraging problem was considered in the previous subsection, here we consider the problem of deblurring using the aforementioned MIM algorithm.

After expanding in the Radon domain, if we assume a deaveraged estimate $\hat{Y}_\theta$ in Eq. (9.16), the problem becomes the following:

$$\text{Solve for } X_\theta \text{ in } Y_\theta = HX_\theta + n, \tag{9.21}$$

where X_θ and Y_θ are the vectors corresponding to the Radon transform at angle θ. The form of Eq. (9.21) is the well-known one-dimensional ill-posed inverse problem over each angle in the Radon domain. We say that the problem is ill posed because the addition of noise in the model precludes the existence of a unique solution. We have noted that the Radon transform of noncursive letters tends to be sparse, or localized. To this end, we propose using the MIM algorithm to do the restoration of each projection in the Radon domain.

9.4.4 Super-resolution and deblurring algorithm

Having described the requisite components, we now present our approach to the SISO super-resolution problem. As we have noted, the procedure closely resembles the IBD alternating minimization approach proposed in Ref. 6. We note that Eq. (9.16) can be decomposed into two ill-posed inverse problems, namely to solve for Y in

$$Y_d = DY \tag{9.22}$$

and to solve for X in

$$Y = HX + n. \tag{9.23}$$

We start by partitioning the text image into characters. While this process is not straightforward, we will assume that the acquired image is of sufficient resolution that this can be done without much difficulty. Now, with regards to Eq.(9.22), we transform the partitioned letter image into the Radon domain using Eq. (9.18). For an $N \times N$ image, this produces a $N \times 180$ Radon image. Next, each column of the Radon image is linearly interpolated by a factor of q corresponding to the desired resolution improvement in the image.

This procedure alone does not produce a valid approximation for the high-resolution blurred scene Y. However, in the next step we deconvolve each column with the MIM algorithm using some guess for σ in the assumed Gaussian channel. The effect of this is to both smooth the low Radon intensity values of the Radon angle vector and to sharpen the peaks in the data. As we have previously mentioned, the MIM algorithm can effectively reproduce these peaks in the data; it constrains the solution set to be nonnegative and the choice of mapping function linearizes the penalty function and has the effect of L_1 regularization. Thus, the deconvolution algorithm used to solve the inverse problem in Eq. (9.23) also refines the solution of the inverse problem in Eq. (9.22).

Next, the expanded Radon domain image is transformed back into the image domain. At this point, the image is sliced so as to enforce the constraint that the image original text image is bimodal. This image, then, is our initial estimate for x. However, this estimate is derived using an uninformed guess for the channel. Therefore, following the IBD framework, we must transform this image into the Radon domain and reapply the MIM deconvolution algorithm, this time solving for the channel H in the Radon domain. After transforming the channel estimate back to the image domain, we may enforce the Gaussian constraint on the channel by performing a nonlinear least-squares Gaussian fit on this channel estimate. Finally, upon convergence, the image block is stored and the channel is used as the prior guess for the next letter image. To reiterate, then, even though linear interpolation is only being performed once, the solutions of both inverse problems in Eqs. (9.22) and (9.23) are iteratively refined through this procedure. The scheme is illustrated in Fig. 9.9.

9.4.5 Results

We test the performance of the proposed algorithm with the images of various Latin letters. The images are blurred by a Gaussian PSF with a variance equal to half the length of the letter. In the original high-resolution scene, the letter image is assumed to be 32×32. After the blurring process, the images were downsampled by a factor of 8. The images were then partitioned to isolate single letters. The results were compared against the output of the IBD algorithm whose input was the spatial domain linear interpolation of the low-resolution blurred image. As expected, degraded images of letters which consist primarily of lines are restored well with this technique. This is shown in Figs. 9.10(a), 9.10(b), and 9.11(a). However, we also see that the technique seems to work equally well with curved letters, as can be seen in Fig. 9.11(b).

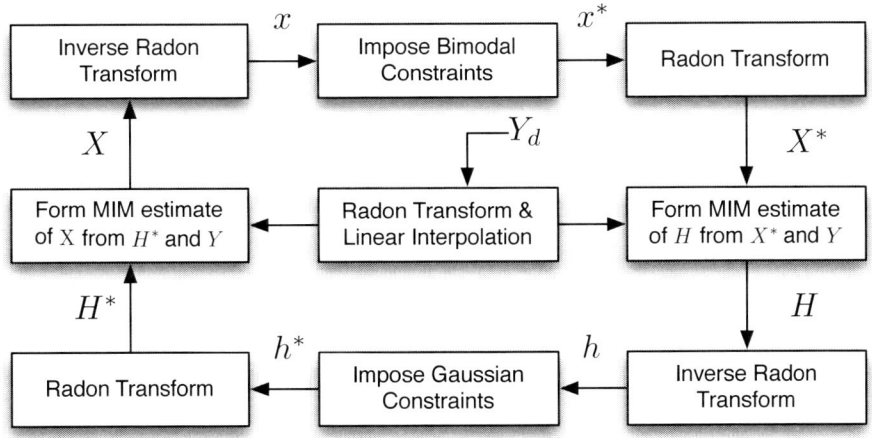

Figure 9.9 Flow diagram of proposed super-resolution scheme

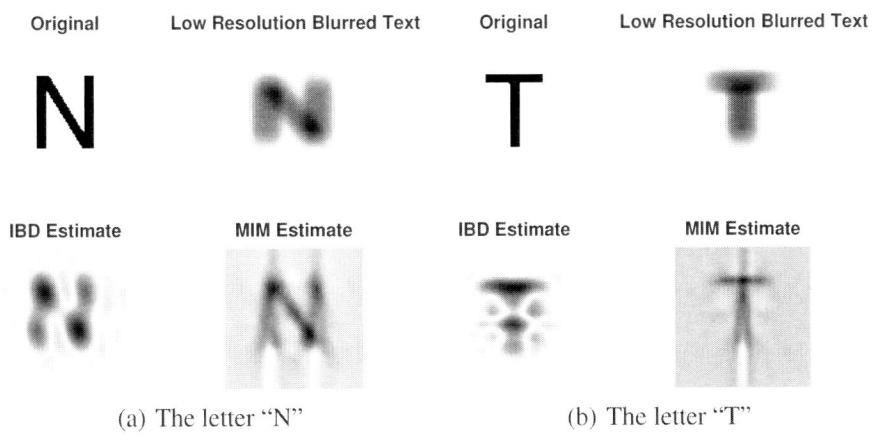

(a) The letter "N" (b) The letter "T"

Figure 9.10 Super-resolution estimates using the MIM and IBD algorithms.

Finally, we point out that the speed of the algorithm can be greatly increased by reducing the number of angles computed in the Radon transform. In Eq. (9.18), θ can be restricted to some smaller set of angles within $[0, \pi)$. Figure 9.12(a) shows the result of applying the algorithm over ten equally spaced points along the interval $[0, \pi)$, while Fig. 9.12(b) show the result of applying the algorithm over twenty equally spaced points. Additional examples and results may be found in Ref. 11.

9.5 Conclusion

An application of the MRE principle to signal-restoration problems was demonstrated. The particular problem dealt with finding a power spectrum given a limited number of noisy autocorrelation function values with known variance. Statistics

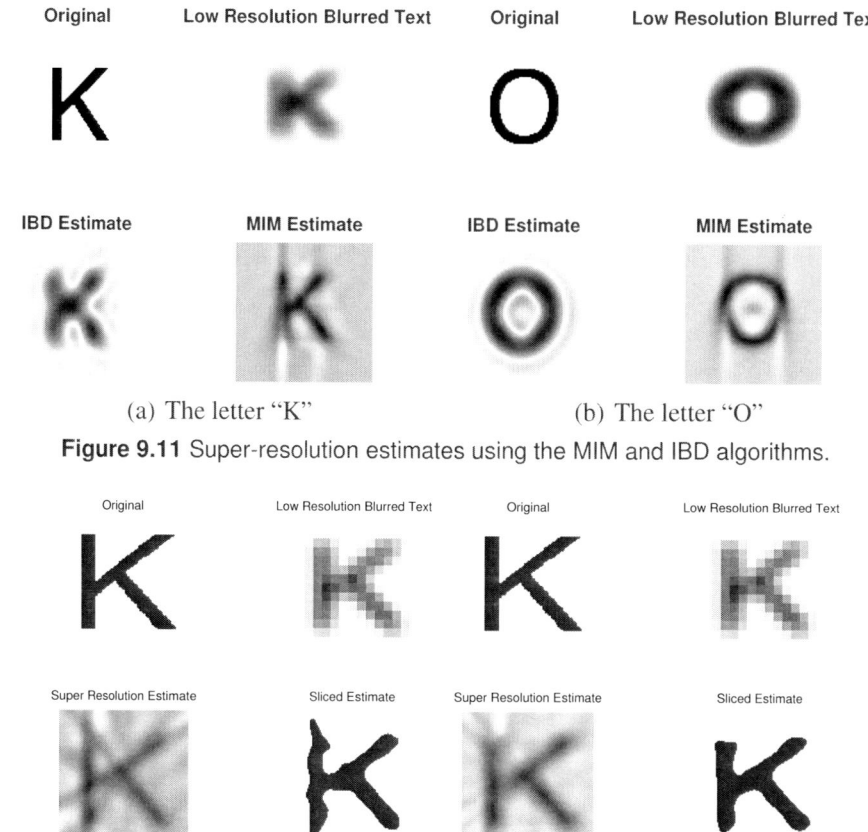

(a) The letter "K"   (b) The letter "O"

Figure 9.11 Super-resolution estimates using the MIM and IBD algorithms.

Figure 9.12 Super-resolution estimates using twenty equally spaced angels (left four figures) and ten equally spaced angles (right four figures).

other than the variance can be utilized by the method. In this case, the form of the final iterative equation will be different.

Next, we proposed a fast alternating minimization algorithm based on the IBD framework that can blindly restore images of noncursive text. By exploiting the properties of the Radon transform, it was shown that a minimum entropy type restoration algorithm could be used to perform both deaveraging and deblurring. Furthermore, it was shown that the algorithm could be easily sped up without sacrificing accuracy.

Finally, we note that only a bimodal measure was used to evaluate the quality of the iterative restoration. Instead, if a combined measure similar to Ref. 10 were used instead, where smoothness and adherence to the low-resolution input are also evaluated, the overall restoration would be better.

References

1. J. Noonan, M. Tzannes, and B. Achour, "Minimum relative entropy enhancement of fft spectra," *Electronics Letters* **26**, 979 (1990).

2. A. Pentland, "A new sense for depth of field," *IEEE Transactions on Pattern Analysis and Machine Intelligence* **9**(4), 523–531 (1987).

3. P. Stoica and P. Babu, "The gaussian data assumption leads to the largest cramér-rao bound [lecture notes]," *Signal Processing Magazine, IEEE* **28**(3), 132–133 (2011).

4. R. Schultz and R. Stevenson, "A bayesian approach to image expansion for improved definition," *IEEE Transactions on Image Processing* **3**(3), 233–242 (1994).

5. D. Rajan and S. Chaudhuri, "Simultaneous estimation of super-resolved scene and depth map from low resolution defocused observations," *IEEE Transactions on Pattern Analysis and Machine Intelligence* **25**(9), 1102–17 (2003).

6. G. Ayers and J. Dainty, "Iterative blind deconvolution method and its applications," *Optical Letters* **13**, 547– (1988).

7. J. Noonan, *The mutual information principle and some of its applications.* PhD thesis, Tufts University (1973).

8. J. Noonan and P. Natarajan, "A general formulation for iterative restoration methods," *IEEE Transactions on Signal Processing* **45**(10), 2590–2593 (1997).

9. B. Achour, *Regularization Theory in Signal Restoration – An Information Theoretic Approach.* PhD thesis, Tufts University (1991).

10. P. Thouin and C. Chang, "A method for restoration of low-resolution document images," *International Journal on Document Analysis and Recognition* **2**(4), 200–210 (2000).

11. P. Basu and J. Noonan, "A new technique for the restoration of low resolution text images," in *IEEE International Conference on Systems, Man and Cybernetics*, (2007).

Appendix A
Generalized Projections

A.1 Popular Classes of Distances

Suppose X is a set with finite cardinality n. We may regard functions over X as finite dimensional vectors, say p and q.

A.1.1 Euclidean distance

The Euclidean distance between two vectors is given by

$$\sqrt{\sum_i (p_i - q_i)^2}, \tag{A.1}$$

and in the continuous case, the distance, called the L_2 distance, is written as

$$\sqrt{\int (p(x) - q(x))^2 \, dx}. \tag{A.2}$$

For nonnegative functions the Euclidean distance is not the best distance measure to choose.[*] Other distances such as the Bregman distance and Csiszár's I-divergence are more suitable.

A.1.2 Two families of distances

Suppose f is a strictly convex, differentiable function on $\mathbb{R}^+$ where $\mathbb{R}^+$ is the non-negative part of the real line. We define the Bregman distance between p and q as

$$B_f(p, q) = \int [f(p(x)) - f(q(x)) - f\prime(q(x))(p(x) - q(x))] \, \lambda(dx), \tag{A.3}$$

and the f-divergence as

$$D_f(p, q) = \int q(x) f\left(\frac{p(x)}{q(x)}\right) \lambda(dx). \tag{A.4}$$

---

[*]The reason for this is the dependence of the Euclidean norm on scaling factors. For example, the distance between x and y is very different from the distance between $2x$ and $2y$. Our interest here is in the shape of the function. For such purposes, this extreme dependence on relative magnitude is not desirable.

It is intuitively pleasing to have a distance that yields 0 as the the distance between two functions that are identical. In the definition above, we assume that $f(1) = 0$ and $f'(1) = 0$.

A.1.3 Examples

- If f is defined to be

$$f(t) = t \cdot log(t) - t + 1,$$

then it follows that $B_f(p, q) = D_f(p, q)$. This is the so called I-divergence measure $D(p \parallel q)$.

- In spectrum estimation the Itakura–Saito distance is a popular measure of the similarity of two estimates. This distance is written as

$$D_{IS}(p, q) = \int \left[-log\frac{p(x)}{q(x)} + \frac{p(x)}{q(x)} - 1 \right] \lambda(dx). \qquad (A.5)$$

Choosing

$$f(t) = -log(t) - t + 1,$$

we see that

$$B_f(p, q) = D_{IS}(p, q),$$

and

$$D_f(p, q) = D(q \parallel p).$$

In a similar manner, one may investigate other forms for f, keeping in mind the requirement that the form be a strictly convex and differentiable function. It is of interest to note here that $B_f(p, 1) = D_f(p, 1) = \int f(p(x)) \lambda(dx)$. This is the so called f-entropy of p.

Next, we discuss generalized projections as applied to the recovery of nonnegative functions in the presence of partial data.

A.2 Generalized Projections

The name *generalized projections* come from the following definition of projection.

Definition A.1 (Projection). A projection of q onto F is that $p^* \in F$, whose distance from q is minimum, provided such a p^* exists.

There exist theorems concerning the existence of a projection, but all of these require sophisticated mathematical techniques beyond the scope of this appendix. In the following, we present some popular estimation problems to which this approach has been applied.

- One example is the problem of inferring a probability density on R when only its first and second moments are known. In addition, the default model may be

$$q(x) = \frac{1}{\sqrt{2 * \pi}} exp\left(\frac{-x^2}{2}\right).$$

(A.6)

In this case, the set of feasible solutions is given by

$$F = \left\{ p \mid \int p(x), \int xp(x) = m_1 \int x^2 p(x) = m_2 \right\}.$$

(A.7)

- Spectrum estimation is another problem where one has to infer a nonnegative function $p(x)$ $0 \le x \le 1$ when, for instance, only

$$\int_0^1 p(x) \cos(2\pi k x)\, dx \quad k = 0, 1, \ldots, n$$

are known, and the default model is $q(x) = 1$.

Appendix B
Definitions from Linear Algebra

It is convenient to represent discrete sequences as vectors (or in the case of two-dimensional sequences, as matrices). This leads to a very compact and convenient notation. A column vector containing N elements is written as

$$\mathbf{u} = \begin{bmatrix} u(0) \\ u(1) \\ u(2) \\ \vdots \\ u(N-1) \end{bmatrix}. \tag{B.1}$$

Throughout this book, all of the vectors used are column vectors. Images are two-dimensional sequences, and it is only natural that we use matrices to represent discretized images. A matrix is written as

$$\mathbf{H} = \begin{bmatrix} h(1,1) & h(1,2) & \dots & h(1,N) \\ h(2,1) & h(2,2) & \dots & h(2,N) \\ \vdots & \vdots & \vdots & \vdots \\ h(N,1) & h(N,2) & \dots & h(N,N) \end{bmatrix}. \tag{B.2}$$

A *Toeplitz* matrix is a matrix whose diagonal elements are identical, and the following equality holds true:

$$h(m,n) = h(i,j),$$

where $m - n = i - j$. Such matrices are commonly used to represent the input-output transformations of one-dimensional linear shift-invariant (LSI) systems.

A matrix whose rows and columns are given by a circular shift of the preceding row or column is called a circulant matrix. Such matrices describe the input-output transformations of linear periodic sequences. Thus, linear convolution can be implemented as a *Toeplitz* matrix operation and circular convolution may be represented as a circulant matrix operation.

A matrix **A** whose elements are matrices themselves is called a block matrix, i.e.,

$$\mathbf{A} = \begin{bmatrix} A_{1,1} & A_{1,2} & \ldots & A_{1,N} \\ A_{2,1} & A_{2,2} & \ldots & A_{2,N} \\ \vdots & \vdots & \vdots & \vdots \\ A_{N,1} & A_{N,2} & \ldots & A_{N,N} \end{bmatrix}. \tag{B.3}$$

If the block structure is *Toeplitz* (circulant), then the matrix is called block *Toeplitz* (block circulant). In addition, if the block elements themselves are *Toeplitz* (circulant), then the block structure is said to be doubly *Toeplitz* (doubly circulant).

It is possible to stack the rows of a matrix one after the other in the form of a vector. Such an ordering is called a row-ordering of the matrix. A similar ordering may be obtained by stacking the columns one on top of the other. The row ordering of a matrix is commonly used in image-processing literature and is referred to as a *lexicographic ordering*. Two-dimensional convolution may be represented as the matrix product of a doubly *Toeplitz* (circulant) matrix and a lexicographically ordered matrix.

Appendix C
Definitions From Analysis

Some of the development in the book uses certain theorems and methods from analysis that the signal-restoration community is becoming increasingly familiar with. In this appendix we provide some of the fundamental definitions and give pointers to references where further details may be obtained.

Definition C.2 (Metric space). A metric space is a pair consisting of an arbitrary set, say $\mathcal{M}$ and a nonnegative real-valued function, say d defined on ordered pairs of elements from $\mathcal{M}$, that is, on $\mathcal{M} \times \mathcal{M}$. The metric space may be denoted by $(\mathcal{M}, d)$ or, when the function d is well fixed in mind, by $\mathcal{M}$.

The function d is assumed to satisfy the following postulates:

- $d(x, y) = 0$ if and only if $x = y$;

- $d(x, y) = d(y, x)$;

- $d(x, y) + d(y, z) \geq d(x, z)$ for all $x, y, z \in \mathcal{M}$;

It is easy to verify that the distance metric* on $\Re^k$ based on the mean-square norm obeys these postulates.

Definition C.3 (ϵ-Disk). Let $(\mathcal{M}, d)$ be a metric space. For each fixed $x \in \mathcal{M}$ and each $\epsilon > 0$, the set

$$D(x, \epsilon) = \{y \in \mathcal{M} \mid d(x, y) < \epsilon\} \tag{C.1}$$

is called the ϵ-Disk about x.

Definition C.4 (Open set). A set $\mathcal{A} \subset \mathcal{M}$ is said to open if for each $x \in \mathcal{A}$, there exists an $\epsilon > 0$ such that $D(x, \epsilon) \subset \mathcal{A}$.

Definition C.5 (Closed set). A set that is not open is a closed set.

Proposition C.1 (Convergence). A sequence x_p in $\mathcal{M}$ converges to $x_* \in \mathcal{M}$ if and only if for every $\epsilon > 0$ there is an N such that $p \geq N$ implies $d(x_*, x_p) < \epsilon$.

---

*By which we actually mean the metric used in this work.

In relation to the analysis of iteration systems, it is essential to know if the iteration will converge to an element of the set over which the iteration system is defined.[†] It is thus necessary to work with sets that contain the points of convergence. A set for which this is true is said to possess the *completeness* property.

Definition C.6 (Completeness). Let $(\mathcal{M}, d)$ be a metric space. A *Cauchy sequence* is a sequence $x_p \in \mathcal{M}$ such that for all $\epsilon > 0$ there is an N such that $p, l \geq N$ implies $d(x_p, x_l) < \epsilon$. The space $\mathcal{M}$ is called *complete* if and only if every *Cauchy Sequence* in $\mathcal{M}$ converges to a point in $\mathcal{M}$.
Finally we state the principle of directed orthogonality.[1,2]

Definition C.7 (Principle of directed orthogonality). A distance or distortion measure Ω is said to obey the directed orthogonality principle with respect to an admissible estimate $\hat{x}$ if

$$\Omega(x, \hat{x}) = \min \Omega(x, y) \ y \in \mathcal{S}$$

is true for all admissible estimates. $\mathcal{S}$ is the set of feasible solutions.

Additional definitions of interest include:

1. Iterative scheme: a multistep procedure that estimates an unknown by successive approximations where the present estimate is a function of the previous one(s).

2. Compact set: a set $A \subseteq M$ is said to be *compact* in the space of real numbers if every sequence $\{x_n\}, (x_n \in A)$ contains a convergent subsequence x_{n_j} with limit $x \in A$.

3. Measurable function: a function is said to be measurable if, for arbitrarily preassigned arguments, there exists a procedure for calculating the value of the function in a finite number of steps.

4. *Nice* function: a continuous, differentiable function with values in the set of real numbers.

5. Best approximation: $\frac{p}{q}$ is called a *best approximation* of x if

$$\left| \frac{a}{b} - x \right| \leq \left| \frac{p}{q} - x \right|, \frac{a}{b} \neq \frac{p}{q} \implies b > q.$$

For an excellent introduction to analysis, see Marsden and Hoffman.[3] Goldstein[4] provides an engineering flavor to the topic by using a constructive approach. The standard text, of course, is Rudin.[5]

[†]Of course, this is relevant only in the case of sequences that converge.

References

1. L. Jones, "Approximation theoretic derivation of logarithmic entropy principles for inverse problems and unique extensions of the maximum entropy method to incorporate prior knowledge," in *SIAM Journal of Applied Math*, **49**, 650–661, SIAM (1989).

2. L. Jones and C. Byrne, "General entropy criteria for inverse problems, with applications to data compression, pattern classification, and cluster analysis," *IEEE Transactions on Information Theory* **36**(1), 23–30 (1990).

3. J. Marsden and M. Hoffman, *Elementary Classical Analysis*, WH Freeman, New York, NY (1993).

4. A. Goldstein, *Constructive Real Analysis*, Harper & Row, New York, NY (1967).

5. W. Rudin, *Principles of Mathematical Analysis*, McGraw-Hill, Inc., New York, NY (1976).

Appendix D
Notation

The following notation is used throughout this work. Any departure from this notation is clearly indicated in the relevant portion of the text.

- Lowercase letters represent signals or images in the time domain. This also holds in the spatial domain, except for i, j, k, l, which are used to represent integers.

- Boldface lower case letters symbolize sampled signals, i.e., vectors. The image matrices are represented by *lexicographically* ordered[*] vectors.

- Uppercase letters stand for the transform domain equivalents of the signals or images represented by the corresponding lowercase letters.

- Boldface uppercase letters are used to denote matrices, for example, the Toeplitz matrix, which is associated with the impulse response of a linear system.

- A hat on top of any symbol indicates that the quantity is an estimate of the associated function.

---

[*]See Appendix B.

Appendix E
Acronyms

AWGN additive white Gaussian noise

CCD charge-coupled device

CLSR constrained least-squares restoration

FFT fast Fourier Transform

GMF generalized mapping function

IBD iterative blind deconvolution

IID independent and identically distributed

KL Kullback–Leibler

LSE least-squares estimate

LSI linear shift-invariant

LTI linear time-invariant

MAP maximum *a posteriori*

MDL minimum description length

ME maximum entropy

MI mutual information

MIM mutual information measure

MIMO multiple-input multiple-output

MIP mutual information principle

MISO multiple-input single-output

ML maximum likelihood

MLE maximum likelihood estimate

MMSE minimum-mean-squared error

MRE minimum relative entropy

PSF point spread function

PDF probability density function

PMF probability mass function

RTSFLS regularization theory stabilizing functional using the least-squares norm

SISO single-input single-output

SNR signal-to-noise ratio

Index

Joseph Noonan received his BS, MS, and PhD degrees in electrical engineering from Tufts University. He began his career with the Raytheon Company in 1970. There he worked on the Patriot and Sparrow Missile Systems. In 1978 he founded Beford Research Associates, a mathematical analysis and scientific software firm that provided support to a number of research facilities of the U.S. Air Force, U.S. Navy, and the Department of Transportation. He served as President of Bedford Research until joining the faculty of the Electrical Engineering Department of Tufts University in 1985. His specialty areas are Statistical Communication and Signal Processing. Dr. Noonan has published over 100 papers in scientific journals and conference proceedings and is the 1992 recipient of the Leibner Award for Outstanding Teaching at Tufts University. He is past chair of the department and is presently Professor of Electrical and Computer Engineering and a consultant to the U.S. Department of Defense. Dr. Noonan is a member of Tau Beta Pi and Eta Kappa Nu.

Prabahan Basu received his BS degree in biomedical engineering with minors in mathematics and electrical engineering from the University of Pennsylvania, Philadelphia, in 1998. He received his MS degree in electrical engineering in 2003 and PhD degree in electrical and computer engineering in 2008 from Tufts University. From 2007–2009, he was a Postdoctoral Fellow at the Statistics and Information Sciences Laboratory at Harvard University. His research interests there included nonstationary signal processing and time-frequency analysis. Currently, he is on the technical staff at MIT Lincoln Labs.